Seeds of Distrust

Nicky Hager

CRAIG POTTON PUBLISHING

For Julia, and her generation, who are told by adults to tell the truth but must learn to live in a world of advertising and PR where it is regarded as legitimate not to.

Published in 2002 by Craig Potton Publishing
98 Vickerman Street, Nelson, New Zealand
www.craigpotton.co.nz

ISBN: 0908802 92 7

Printed by Astra Print, Wellington, New Zealand

CONTENTS

ACKNOWLEDGEMENTS

The story in this book had been carefully hidden and could easily have stayed that way. The most important step to uncovering it was a chance remark from someone involved who knew there was much more to the issue than the public had been told. Since then I have received assistance from other people who knew parts of the story and helpfully provided documentation. I am grateful to them for agreeing to help and for their trust in me.

As with all my work, I have relied a lot on my friends and family for practical assistance during the research and writing. I have again enjoyed working with the people at Craig Potton Publishing and Astra Print, especially Robbie Burton, and also editor, James Brown. Thanks to all these people.

FOREWORD

In 1999 I co-authored a book about the relationship between government, business and public relations companies. Our conclusion was that secrecy allowed the then National Government to abuse power as it allowed state agencies to mislead the public and play politics in pursuit of their preferred policies. After nine years in office, National had become arrogant and complacent in its exercise of the power that citizens had conferred on it.

Three years later there is a different government which, on various constitutional and democratic issues, has behaved with greater integrity. However, on issues of secrecy, susceptibility to business interests and misuse of executive power, some of the same issues arise.

This book is a case study of how the Labour-Alliance Government handled an issue they would have preferred remain secret, and how they made decisions of which they were not wholly proud. As a case study, I am not suggesting that this is typical of all or even most decisions. But it is precisely in the case of hard issues, controversial issues and issues where there may be partisan advantage in acting questionably, that democratic checks and accountability are most important. Unless governments are challenged, they drift towards the complacent exercise of power that characterised the nine-year-old National Government in 1999.

The controversial subject of my 1999 book, *Secrets and Lies,* was logging of New Zealand's native forests, which at that time had become a bitter political conflict. The present book is about an equally controversial subject: genetic engineering.

I should state clearly that I am not claiming to have anything expert to say about genetic engineering. That issue is the context rather than the subject of the book. This story is about the importance of freedom of information, public accountability and the great disease of modern politics: the assumption that it is legitimate for those with power to manage and manipulate the news.

I will also state my personal view about genetic engineering, which is mixed. I do not assume that all gene technology is bad. I see it as being like New Zealand's 'nuclear free' policy, which rejects the needlessly risky and completely immoral sides of nuclear technology but does not stop nuclear medicine in hospitals. With any major and fundamental new types of technology, there are huge potential risks. The problem with genetic engineering is that if something goes wrong, it is likely to be the result of some surprising interaction of factors or some unforeseen biochemical effect that science would only properly understand in retrospect. It could take generations to realise what had happened and why.

My personal uneasiness with the technology stems from the fact that a lot of genetic engineering is being done by companies in a hurry to cash in on commercial applications, and hurried science is not always good (or safe) science. Thus the key issues are caution and the degree to which decision-makers focus on the public interest as opposed to short-term gain. With a lot of genetic engineering being driven by a desire for short-term profit, I believe the public is justified in feeling concern and even distrust about what unforeseen costs may fall upon ordinary people and the environment. Trust and responsibility are the central themes of this book

The difficult thing for politicians in government is that they have, at the same time, two separate and often conflicting roles. Most visible are their party political roles, where they are trying to stay in power, look good to the public and implement their policies. But most important are their formal ministerial roles as players within and protectors of constitutional processes.

One of the fundamental constitutional issues is freedom of information. Huge political advantage can be gained by a government through deciding what information is released on an issue, how it is released or even whether it is released at all. But, at the same time, the right to information – the public's ability to know what is going on – is the foundation of democratic government.

We have become very used to secrecy, both from government and business, and the public relations tactics of selectively releasing information in ways that suit those in power. We are often left with the feeling that, if not actually being lied to, we are not getting the whole story. My view is that although we are used to it we should not accept it as right. I believe that government information does not belong to the Government of the day, to do with it what it likes. It is the public's information about the public's business, which we have entrusted to a Government to manage for us. As such, the secrecy and public relations 'spinning' of information should be viewed as a fundamental abuse of government power.

To be fair to ministers and government departments, most of the time on most subjects they quietly and faithfully fulfil their public duties. It is on controversial issues, issues that could be embarrassing and subjects where politicians or their allies stand to gain or lose advantage, that abuse of power occurs.

Besides trust and the freedom of information, the other theme of this book is the role of private public relations companies in helping their corporate clients influence government. Most ministers will say that industry PR people might help their clients have their say to Government, but that they do not give special influence to business over citizens. I think they are wrong.

My impression is that on a controversial issue like genetic engineering – where huge companies, world-wide agendas and lots of money are involved – the public relations strategies of the big players often have more influence on government decisions than public opinion, science or the public interest. That is the struggle of democracy in a business-dominated world.

Even the names people use for the technology have been 'spun'. The original name is 'genetic engineering' (GE), referring to techniques of shifting around or removing genes in organisms' cells, or making 'transgenic' shifts of genes between different types of organisms. As public suspicion grew towards some of the technology, US companies like Monsanto started to use the new, blander term 'genetic modification' (GM). Some US food companies have since moved to stay ahead of public concerns and now use the term 'genetic enhancement'. For this book I will use the original name adopted by genetic scientists, 'genetic engineering'.

This book is not intended as a personal criticism of the individuals involved, and so I have mentioned a minimal number of people by name. Some people deserve to do some soul searching, but most of the people in the story (including those named) were not acting unethically. They were just busily following instructions from their superiors. The book is about the systemic problems that lead people in authority to make socially irresponsible decisions.

At various points in the book I have highlighted questions which I hope the companies and individuals involved will answer publicly. I would like to have sought their views to include in the book. The reason I did not was that I wanted to avoid possible publication delays by those who might prefer the revelations did not occur. Even when the contents are factual and fair, it is relatively easy to tie up a book in the courts. Therefore I was advised to avoid pre-publicity. However, I have sought to ensure accuracy by talking to a range of insiders, confirming details with outside experts and by relying heavily on documentary evidence. Also, I have included a set of the Labour-Alliance Government's own papers at the end of the book – Cabinet documents, media materials and correspondence – so that readers can judge them for themselves.

I considered carefully whether to release the book before the New Zealand 2002 general election. An election is the natural time to raise issues of accountability and a Government's record. I decided that election year was good but rushed to complete the book well before the campaign period. When the early election was called, and genetic engineering had become an election issue, I seriously reconsidered whether it was appropriate to release the book in the midst of political campaigning. In the end I decided that, as is the theme generally in this book, the public had a right to know.

WHO'S WHO

CABINET MINISTERS

Helen Clark, Prime Minister

Marian Hobbs, Minister for the Environment and of Biosecurity

Pete Hodgson, Acting Minister for the Environment

COMPANIES

Novartis Seeds (now Syngenta), the multinational seed company that supplied the contaminated sweet corn seeds.

Cedenco Foods Ltd, which grew GE contaminated crops around Gisborne

Heinz Wattie Ltd, which grew GE contaminated crops in Hawkes Bay

Talley's Frozen Foods Ltd, which grew GE contaminated crops near Blenheim

DPMC **Department of the Prime Minister and Cabinet**, where all day-to-day management of the contaminated crop issue occurred.

ERMA **Environmental Risk Management Agency,** the agency that decides on applications to introduce new organisms (including GE plants) into New Zealand.

LABORATORIES

Crop and Food Research, Lincoln

Genescan, Melbourne

LSN **Life Sciences Network**, the pro-GE industry lobby group.

MAF **Ministry of Agriculture and Forestry**, which oversees New Zealand's bio-security laws and provided advice on the contaminated sweet corn issue.

MFE **Ministry for the Environment**, the agency that coordinated policy advice on the contaminated sweet corn issue.

MFAT **Ministry of Foreign Affairs and Trade**, which provided advice on the World Trade Organisation and free trade implications of trying to stop GE contaminated seed imports.

NGA KAIHAUTU TIKANGA TAIAO
The ERMA Maori advisory group, which is consulted on the Maori aspects of ERMA decision making.

PR COMPANIES
Allan and Boyle, which represented Cedenco Foods Ltd.
Communications Trumps, which represented Novartis Seeds and took the lead role in negotiations with Government.
Hill and Knowlton, which represented Heinz Wattie Ltd.

ROYAL COMMISSION ON GENETIC ENGINEERING
The New Zealand royal commission that recommended in 2001 that New Zealand should 'keep its options open' by continuing with research into and eventual release of genetically engineered organisms.

CHAPTER 1

NINE FOOT TALL WARRIORS

The pro-genetic engineering 'Life Sciences Network' opened its presentation to the New Zealand Royal Commission on Genetic Modification on Wednesday, 1 November 2000, with an attack on its opponents. Network chair, William Rolleston, said the organisation had been formed to counter 'uninformed and unbalanced' public debate where science was being 'misinterpreted and ignored' by anti-genetic engineering groups and the media. He argued that 'economic arguments are persuasive for the acceleration of our effort in the field of GE, and not for holding back.'[1]

The Life Sciences Network had formed six months earlier and hired offices in the same building as the Royal Commission on Genetic Modification from which to run its campaign. Headed by a public relations consultant and supported by a large budget from undisclosed sources, the Network has arranged and co-ordinated most of the political pressure in favour of genetic engineering in New Zealand since.

The presentations that Wednesday were confident – all about how, in Rolleston's words, 'responsible scientists and responsible businesses' were using genetic technology for the benefit of New Zealanders. Safety, predictability and caution were the themes of the day. Any public concerns, he explained, were the result of lack of understanding of science or misleading statements from environmental groups. It was simply a case of 'real science' versus 'junk science'.

The Network had hired a Maori consultancy firm to open the day with a karakia and then explain why Maori opposition to genetic engineering was an unsophisticated 'formative reaction'. The consultants argued that historically Maori were quick to adopt new technologies, which is true, but also that 'Maori were practising genetic engineering long before settlement in this country'. They argued that in Maori mythology the decision by Tane and his brothers to separate the god parents Rangi (sky) and Papa (earth), from one into two, was 'a nice sort of metaphor ... perhaps the very first in the Maori conceptualisation, [for] genetic modification.'

The hired consultants spoke of stories in Maori history about nine-foot-tall warriors. 'Maoridom, I believe, were already in the processes of reproducing warrior machines in their history.... If they were nine feet tall, there must have been some genetic modification in the processes to create such a being and such a warrior.'[2]

The next Life Sciences witness, flown from Switzerland for the hearing, was ecologist Klaus Amman, a practised advocate of genetic engineering. With a large grey beard and attractively imprecise English, he flicked rapidly through slides – 'Here, nice graphs, no time to explain properly...' – and had the Commissioners laughing at a joke about how, if he were an insect, 'to be honest I would like to be on holiday doing organic farming'.[3]

The real problem in the GE debate, he said, 'that I have to explain here clearly and bluntly', is the 'powerful protest corporations' that 'often are not interested in scientific facts any more'. He said that while some other industries had hidden worrying facts to avoid a public backlash over their activities, 'I don't think you can put the genetic engineering industry in this position'.

> *The nuclear industry hid back a lot of evidence, the tobacco industry did, the chemical industry did, but the biotech industry*

> *never ever held back any negative evidence of genetic engineered crops. That you cannot bring up, and if you can bring it up I'm getting very mad at these people and I'm the first to advocate it.*[4]

By strange coincidence it was on the very day of this hearing – 1 November 2000 – that the first serious incident concerning genetically engineered crops in New Zealand was discovered. Despite Professor Amman's confidence in the genetic engineering industry, the story was hidden and – until the publication of this book – remained hidden from the New Zealand public.

The story began when, on that Wednesday, routine testing of a large batch of sweet corn seeds being imported from the United States indicated they were contaminated with genetically engineered seeds. Most seeds entering New Zealand at that time were not tested at all, but because in this instance some of the sweet corn crops were being grown for a company that guaranteed its produce was 'GE-free', that company had introduced testing to ensure no genetically engineered seeds were used.

By the time the test results arrived, many thousands of genetically engineered seeds had already been planted out and the GE corn plants were quietly growing in the midst of non-GE crops in three different parts of the country. In several weeks the GE plants would flower, spreading their pollen to the surrounding conventional crops and, not long after that, the GE corn crop would be appearing ready to eat in shops.

It could not have happened at a worse time for the genetic engineering companies. Only two months earlier news had gone around the world about the 'Starlink' GE corn strain. Starlink corn, grown in the United States as stock food but not approved for human consumption, had been found mixed in with approved food quality corn supplies. The GE-contaminated corn was first found in taco shells on sale in US shops, then in many other food brands, and a few weeks later in exports arriving in Japan. The Starlink corn contamination had not only been highly controversial, damaging US public confidence in the big pro-GE US food companies, but had also cost the companies concerned tens of millions of dollars as they recalled products from shop shelves.

In New Zealand, just the year before, there had been controversy over another kind of genetically engineered food. Leaked public relations plans about genetically engineered salmon had shown the PR firm Communications Trumps advising that possible escapes into the wild and problems in the experiments should be kept secret from the public. Public trust had been undermined by revelations that Communications Trumps had said 'Issues such as deformities, lumps on heads etc should not be mentioned at any point to anyone outside'. The reasonable public reaction was to question why they should trust assurances about the safety and careful control of genetic engineering when they saw a company hiding things it did not want the public to know.

Worst of all for the pro-GE lobby, this first (albeit accidental) major release of genetically engineered plants into the New Zealand environment – the unapproved, uncontrolled and probably unlawful release of GE sweet corn – had occurred right in the middle of a Royal Commission that would make recommendations about the future of genetic engineering in New Zealand. The Commission was still receiving public submissions when evidence of the contamination appeared. Neither the Government nor the companies concerned wanted anything to stir up public anxiety about and opposition towards genetic engineering at that time.

The sensible course of action was obvious. The crops could have been pulled out and the rest of the contaminated seeds recalled. The public could have been informed of the accidental release and of the action taken. Such a strategy could have built public trust in the willingness and ability of government and business to responsibly handle the issues of genetic engineering.

None of these things happened. There was a more tempting option. Secrecy. Company staff and government officials were sworn to secrecy, even Ministers were told that no leaks would be tolerated. Almost every part of the story in the coming chapters was kept secret from the public.

Seed contamination is not a small side issue in the debates over genetic engineering. It is currently the main way that uncontrolled genetically engineered organisms are being spread around the world. Before we get into the New Zealand story it is worth looking overseas, where GE contaminated crops have been appearing and where other governments have been reacting quite differently to the problem.

Large-scale commercial use of GE crops began in a few countries in about 1996. At that time, much of the public debate centred around whether the GE crops would spread to contaminate conventional crops. The companies vigorously dismissed these concerns. Gradually, though, more and more evidence appeared suggesting that this was exactly what was happening. In many cases it was left to public interest organisations like Greenpeace to uncover the evidence.

In 1998, for instance, Greenpeace arranged for GE testing of corn cobs from a farmer in southern Germany whose crops were growing next door to a crop of Novartis GE corn, the same company as in the New Zealand story. The samples were analysed at the GeneScan laboratory in Freiburg. The tests found that the Novartis GE corn had cross-pollinated with the neighbouring conventional corn, affecting 5% of the plants at the field border and 1% at 10 m inside the conventional crop. Neither Novartis nor the German authorities had warned neighbouring farmers of the GE plantings.[5]

In 1999, as customer resistance to GE food grew in Europe, a Dutch food importer found that supposedly organic tortilla corn chips from the United States contained GE corn and refused to buy them for the premium organic produce prices. The Wisconsin organic food company, Terra Prima Inc., said it had to assume the contamination came from GE pollen because it had not been told about any nearby GE crops. Numerous cases like these appeared at that time, raising concerns among conventional and organic farmers in countries with GE crops growing that their businesses were at risk.

Jeff Stein, director of regulatory and government affairs for Novartis, was quoted responding to the Terra Prima case. He said

Austria, like New Zealand, prides itself on its high environmental standards. In June 2001, a few months after the events in this book, the Austrian authorities discovered GE contaminated corn crops growing on Austrian farms. Unlike New Zealand, they paid their farmers to pull them out. (Photo: TRANZ/Corbis)

'organic farmers could reduce the risk of contamination if they follow standard farming practices like creating a buffer between neighboring fields and planting at different times than their neighbors.' In other words, it was not the GE seed companies' responsibility. 'If an organic farmer wishes to ensure no corn pollen in his fields, he needs to take the necessary precautions', he continued.[6] This was assuming, of course, that farmers even knew about the neighbouring GE crops.

During the northern hemisphere growing season in mid-2000,

shortly before the New Zealand crop contamination, more and more cases of seed contamination were appearing. The more stories appeared in the news, the more nervous ordinary people became. Most shoppers are not experts in biotechnology and have to trust companies and governments to look after their health. If they start to doubt that companies and authorities are being suitably careful, they become more cautious about what they buy. In many countries, public resistance to GE food grew very quickly and profit-conscious food companies and supermarket chains responded rapidly, insisting that their suppliers provide non-GE produce and advertising GE-free policies.

But when it came to GE seed contamination, it was quite different. The same GE corporations and lobby organisations that had argued for years that GE crops would not contaminate conventional crops, suddenly changed their tune. Now they argued that not only was GE contamination of conventional seed supplies widespread, it was 'inevitable'.

The problem they had been denying was suddenly presented as the status quo that the world would just have live with. The companies were not responsible, they did not have to act to avoid the problem and, conveniently for them, the controversial line between conventional and GE crops was becoming more and more blurred. Instead of acting to stop seed contamination, they tried to ensure that governments (like New Zealand's) would not introduce rules banning contaminated seed imports. This is the context in which the New Zealand story occurred.

Other governments responded to the new problem of GE contaminated crops quite differently to New Zealand's. One example is Austria's. Like New Zealand, Austria prides itself in having good environmental standards. In April 2001, shortly after the events described in this book, Austrian Greenpeace had two sets of independent GE tests conducted on samples of corn seed imported from the United States. A large batch of conventional Pioneer corn,

PR39D81, was found to be contaminated with two forms of GE corn: the same Novartis Bt11 corn as suspected in the New Zealand contamination, and Monsanto MON810 corn. The corn was being sold by a chain of shops called Lagerhaus.

After Greenpeace published the test results, the Austrian authorities did not initially react and the seed continued to be sold. On 2 May 2001, Pioneer put out a press release saying that the PR39D81 batch was not contaminated. However, behind the scenes the authorities had been doing their own tests and discovered that 'significant' contamination was present. Like the New Zealand tests, the results showed uneven contamination concentrations, with many results registering zero or under 0.1%, some around 0.1% and some higher.

In June, the Austrian regional authorities announced that Pioneer must withdraw the contaminated seeds from shops and asked farmers to destroy the contaminated crops on a voluntary basis. The regional authorities would then compensate them for the lost crops and Pioneer would in turn compensate the authorities. It was a very different attitude to that of New Zealand.

In total 180 tonnes of contaminated seed had been sold. The Austrian authorities managed to locate 530 affected farmers and paid a total of 2.76 million euros compensation. Pioneer refused to concede any liability and in late 2001 had still not paid any compensation.[7] Since this time Austria has strictly tested seed imports, and backed this up with tests on the resulting local harvests, showing that it is possible for a government to be serious about GE crop contamination.

The same issues arose in Greece, six months before the New Zealand case. In mid-2000 about 14,000 tonnes of cotton seed was planted in Greece, mostly in the Trikila region, and then found to be contaminated with a GE cotton variety. No genetically engineered cotton is authorised in the European Union. It was the third case of GE contaminated seed to hit the headlines in Europe that summer, the others being contaminated rape-seed and contaminated corn.

The Greek Vice Minister for Environment, Elias Efthimiopoulos, announced on 20 July 2000 that all the contaminated cotton crops had to be destroyed. His Ministry issued instructions that where possible this be done before the plants flowered.

It could easily have been the same in New Zealand.

CHAPTER 2

SOWING THE SEEDS

The contaminated consignment of sweet corn seed arrived by ship from the United States on 10 October 2000. It had been harvested several weeks earlier, at the end of the northern summer, in the highly irrigated Snake River region of southern Idaho in the dry northwest United States. The problem is that about 40% of the corn grown in the United States today is genetically engineered. So – as the Starlink corn controversy had shown graphically only a few months earlier – it is possible for GE pollen to drift from GE corn crops and contaminate ordinary corn, and/or for the GE and non-GE seeds to get mixed during harvesting, processing and distribution.

The corn was supplied by Novartis Seeds. Novartis, now called Syngenta, is one of the multinational GE giants. The Swiss-headquartered corporation develops new varieties of sweet corn in a research subsidiary in Boise, Ihado, and sells sweet corn seeds to farmers around the world to grow for their local markets. Novartis had harvested a 20 tonne batch of this 'Jubilee' sweet corn 'Lot NC9114', of which 5.6 tonnes was sent to New Zealand. Owing to concerns about genetic engineering in Europe, Novartis had recently introduced routine testing of its seeds for GE contamination, but when Lot NC9114 was tested on 8 August the results had been negative.

The 5.6 tonnes imported into New Zealand amount to about 40 million seeds. The shipment, once all used, represented about 7%

(400 ha out of 6000 ha) of New Zealand's annual sweet corn crop. Lot NC9114 seeds were supplied to three companies – Heinz Wattie, Cedenco and Talley's for immediate planting, and to the South Canterbury distributor Seed Production (NZ) Ltd.

By the beginning of November, 178 ha of the seeds had been planted out in three regions of New Zealand. Around Gisborne, about 90 ha had already been planted by the food processing company Cedenco, with the crop due to become dried corn powder exports to Japan and frozen sweet corn supplied, under a five year contract, to Heinz Wattie for sale to New Zealanders and export markets.

The young sweet corn plants were already about 30 cm high when Cedenco got back the positive test result for the 'nos' gene sequence commonly used as a marker in genetic engineering. The laboratory estimated about 0.5% GE contamination of the seeds. Cedenco would have realised the seriousness of the situation: under New Zealand law it is illegal to import or possess an unapproved new organism (viable GE seeds are 'new' plants), plus it had made undertakings to provide 'GE-free' produce.

The day after the positive test result arrived, Cedenco contacted the regional office of Novartis in Melbourne to question it about the seed batch. Cedenco and Novartis agreed to retest the seeds for genetic engineering content. Novartis also contacted the other New Zealand customers that had received the contaminated seeds and identified where they had already been planted. All the companies agreed to cease planting and hold their stocks of seeds until they had worked out what to do next.

Novartis contacted the Ministry of Agriculture and Forestry's Biosecurity Authority to report the incident. MAF took separate samples of the seeds, which were then stored away in case they were needed as evidence in a court case. Seed samples were sent for GE testing to labs in New Zealand, Australia and the United States. All this was underway when officials first informed the Minister for Biosecurity, Marian Hobbs, on Monday, 13 November 2000.

Tests in November 2000 found that thousands of genetically engineered sweet corn plants were growing amidst ordinary sweet corn crops in three regions of the country (including here, near Seddon, in Marlborough). The 5.6 tonne batch of GE contaminated seeds was supplied to Cedenco Foods, Heinz Wattie, Talley's and Seed Production NZ Ltd.

MAF and the companies agreed that the whole incident should be kept secret while they investigated it. According to information compiled by MAF in those first days, a little under half of the contaminated seeds had been sown by the time the planting was halted.

The plantings were:

Gisborne Area: 11 growers covering 93.7 ha (1257 kg seeds)

Hawkes Bay Area: growers in Clive, Hastings and Meanee covering 27.7 ha (381 kg seeds)

Marlborough Area: 4 growers around Seddon covering 57 ha (845 kg seeds).

Besides the unplanted seed held in these areas, another 1000 kg of seed was being held by Seed Production (NZ) Ltd in its warehouse near the port in Fraser Street, Timaru, ready for sale to sweet corn

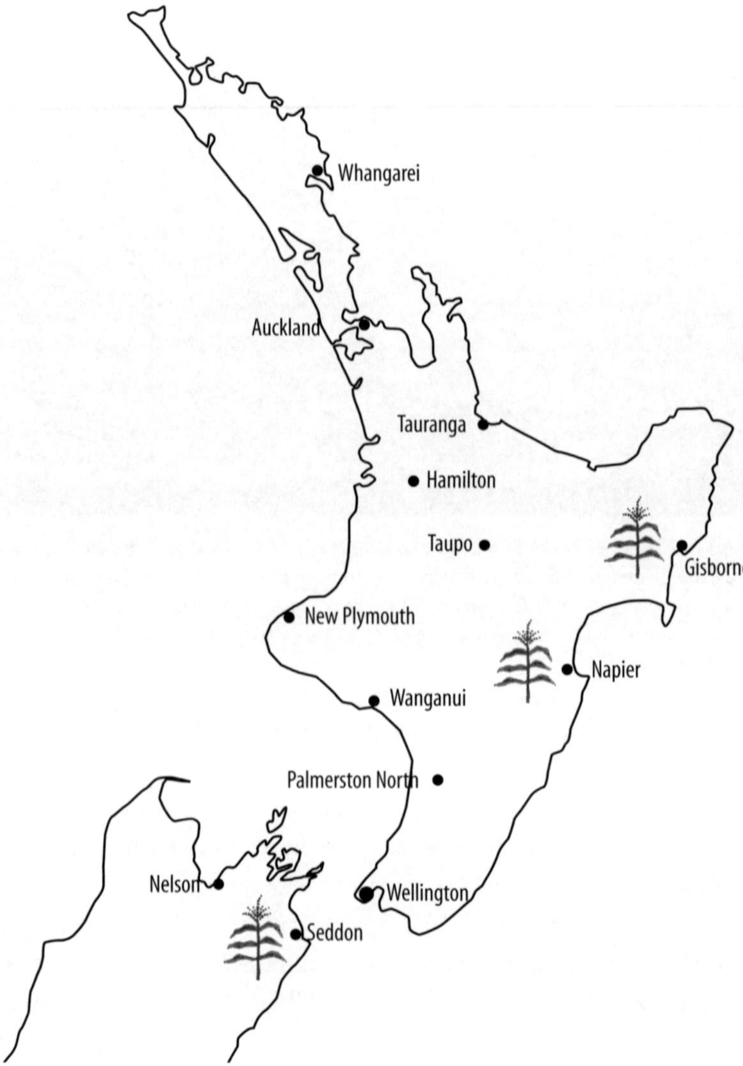

178 hectares had been planted around Gisborne, Hawkes Bay and Blenheim when the GE contamination was detected. In total, the seed batch accounted for 7% of sweet corn production in New Zealand that year. Initially, Helen Clark's 'bottom line' was that the crops must be removed, but in the face of lobbying by industry and officials her bottom line faded.

farmers in that region.

Genetically engineered plants and animals are covered by the same law that prohibits or controls organisms like snakes and new insects being brought into the country: the Hazardous Substances and New Organisms Act. This requires any new organisms being brought into New Zealand to be approved by the Environmental Risk Management Authority (ERMA).

It is a strict law and for good reason. Once a new organism (plant, animal, bacteria etc) is released into the environment, there may be no way to get rid of it again. The possums and destructive weeds that damage our native forests, and the ever-present risk of poisonous spiders and other insect pests arriving unnoticed in shipping containers are all testament to the importance of New Zealand's strict biosecurity laws. Likewise, this is why the idea of genetically engineered 'super-salmon' escaping into our waterways and potentially cleaning out other fish species, caused such concern to both the public and the PR company handling the issue on behalf of King Salmon. When a new organism is released into the environment, things can go very wrong.

The law appears to apply equally to GE sweet corn plants. It says clearly that 'No new organism shall be imported, developed, field tested, or released – otherwise than in accordance with an approval issued under this Act.' Reasonably, you cannot be convicted if you import a new organism accidentally; it must be 'knowingly'. But you are still committing an offence if you then 'possess' a new organism 'imported, manufactured, developed, or released in contravention of this Act'.

It appears, then, that all the companies in possession of the contaminated sweet corn seeds were at risk of seriously breaking the law, a law that provided for fines up to $500,000, possible imprisonment, destruction of the crops and remaining seeds, and an extra $50,000 fine for every day that the offence continued.[8] They were also breaking the moratorium that had been imposed during

Corn field and Young Nicks Head, near Gisborne. The GE contaminated corn crops around Gisborne were growing next to other conventional corn crops and even one organic sweet corn crop. But the other farmers were not told.

the Royal Commission on Genetic Modification, which had expressly halted the release of GE organisms while the Royal Commission looked at the issues.[9] The companies explained to the government authorities that the GE seeds had come in without their knowledge and that they were anxious to co-operate fully in dealing with the accidental release.

But does one release of GE sweet corn really matter that much? After all, there have been undisclosed genetically engineered ingredients in our food in recent years (including ones made from GE corn). Releasing living organisms into the environment is, however, fundamentally different to eating processed food. Ten years from now New Zealand might discover that this or that GE ingredient is not as safe as first thought and recall it. But by the time you find out the bad news about a plant that has been released into the environment (or one whose genetic material has transferred to another species), there may be no way to stop it.

With these concerns in mind, consider how the Environmental Risk Management Authority treated an application for a field trial of genetically engineered corn just 12 months before Lot NC9114 arrived in New Zealand.

In field trials approved under the Hazardous Substances and New Organisms Act, a GE crop can be planted for experimental purposes provided it is strictly contained to ensure neither the plant nor its pollen escapes into the wild. The Authority can also require that all plant material is re-exported or destroyed at the end of the trial.

A 1999 application for GE corn field trials (which were approved but did not end up happening) was made by the international GE seed company Pioneer. Two varieties were to be trialled: Monsanto 'Roundup Ready' corn (where the fields can be sprayed with the Monsanto herbicide Roundup without hurting the crop) and insect resistant 'Bt corn'. This application is especially relevant because the contamination in Lot NC9114 was subsequently thought to be genetically engineered Bt corn.

The rules set by the Environmental Risk Management Authority for the trial sought to ensure 'containment' by requiring all the transgenic corn ears to be individually covered with waterproof paper bags to avoid pollen escaping. Pollination was to be done by hand, transferring the paper bag from over one ear to another and putting a new bag over the first. In case any pollen escaped, the field trials had to be at least 400 m from other corn crops in the vicinity. At the end of the hand-pollination, all the transgenic plants were to be 'destroyed on site by mulching and burial'.[10] Examples of the paper pollen bags were handed around the room during the hearing of this application to explain why they would be reliable. All that trouble for a 0.2 ha trial to ensure no pollen could escape.

In its decision, the Authority acknowledged concerns that had been raised about the Pioneer trial. It noted Maori concerns about transgenic pollen crossing with varieties of corn 'used by Maori for producing the food kanga wai, kanga kopuwai and kanga pirau (fermented corn)'. It also noted 'the uncertainty of genetic engineering as a science', the possibility of 'long term adverse effects materialising well into the future' and the possibility that escaped seeds or pollen 'would jeopardise New Zealand's expanding organic agricultural and horticultural industries, its "clean green" image and its biodiversity'.

'The Authority does not dismiss any of the concerns expressed', it concluded, but these issues would not arise 'while the field trials are undertaken in containment'. It said that the concerns about 'scientific uncertainty and potential long term adverse effects' are 'more relevant to release applications than to an application for small scale contained field tests.... The caution required of the Committee relates to the adequacy of the containment conditions and management regime.'[11]

Now here, a year later, transgenic corn was released and growing freely in Gisborne and Hawkes Bay, centres of Maori life in New Zealand, scattered in the midst of other crops which, together with the GE plants, might soon become food for Maori and other New Zealanders. There was, of course, no containment. There was no control of pollen. No pollen bags. There were no special buffer zones between the GE crops and ordinary crops. There was nothing to stop bees collecting the GE pollen.[12]

All of this helps explain why in early November 2000, when the contamination was discovered, the companies offered to voluntarily remove the crops should the Government require them to.

But, as the following chapters document, over the next four weeks the companies and Government gradually shifted from 'do it properly' to 'cover it up' mode.

In the end, all the Lot NC9114 corn was left to grow, so that by Christmas it was flowering and spreading its pollen to surrounding corn crops. Not only that, government approval was given for the

iii. the ease of eradication of any population established.

The ability of the organism to escape from containment

In considering the ability of the organism to escape from containment, the Committee considered the following specific points:

i. **breach of containment through dispersal of pollen**

Maize is a wind pollinated crop and the pollen is not known to be dispersed by insects. The applications state that the pollen remains viable for a period of up to 30 minutes and the crop will be at least 400m from other maize crops in the vicinity of these trials, which is double the OECD required isolation distance to achieve variety purity.

The applicant has proposed and this decision requires that the pollen tassels be secured in weatherproof paper bags prior to pollen formation. The collected pollen is transferred to the recipient non transgenic maize and secured with the paper bag. New bags are secured over the transgenic donor plants to ensure no remaining pollen escapes.

The Committee considered that should pollen bags be damaged and pollen dispersed from the site, the 400m isolation distance (minimum) from other maize crops, the presence of a shelter belt, and the limited period of viability of the pollen would make it highly unlikely that any pollen could successfully pollinate maize outside the trial site.

ii. **breach of containment through dispersal of seed**

All transgenic maize ears will be contained within the weatherproof paper bags, which are only removed at harvest. This ensures that the transgenic ears are readily identifiable and makes it

GE corn field trial rules – When a small field trial of GE corn (which did not proceed) was approved the year before, strict rules were imposed to stop pollen escaping. Now thousands of GE corn plants were growing, completely uncontrolled, in the environment.

unused seeds to be planted. This included the 1000 kg of unused seed in the Timaru warehouse – which had initially had its biosecurity clearance withheld. Was this also planted and later harvested and sold to the public?

This means permission was effectively given for the 'knowing' release of thousands of GE corn plants, without the approvals required under law, during a full moratorium on GE releases and in the middle of the Royal Commission on Genetic Modification.

I found no evidence that any of the government agencies bothered to do field checks or any other form of monitoring to find out what the effect of this release into the New Zealand environment has been. Presumably no one involved wanted to know.

We have no way of knowing whether eating that particular type of genetically engineered corn (whatever it was) poses a health risk. Lots of Americans eat GE corn all the time and the Australia New Zealand Food Authority had, in 2000, given temporary approval to imported foods containing some GE corn ingredients. But the critical issue is that New Zealanders were not given a choice over whether or not to buy and eat it. This was the whole point of New Zealand introducing a GE food-labelling regime. Even assuming a low level of contamination, 400 ha of contaminated corn amounts to about five tonnes of GE sweet corn, the equivalent of some 10,000 cans of genetically engineered corn.

More immediately affected were the people living and growing their own crops next door to the contaminated fields. Some of the Lot NC9114 crops grown in the Gisborne area were adjacent to other corn crops. None of the neighbours of the contaminated crops were informed either.

The issue of responsibility to one's neighbours is particularly acute in the case of organic farmers. At least one of the crops of contaminated seed was growing across the fence from an organic farm's organic sweet corn crop. The organic farmer risked losing organic certification and suffering financial loss if, as was perfectly possible, pollen drifted over the fence resulting in hybrid genetically engineered sweet corn.

This book is primarily about the politics rather than the details of the contaminated seed story: understanding why these decisions were made, and the bureaucratic and scientific manoeuvring used to justify them. The key issues are constitutional and democratic.

However, before we move on, let us note some questions about the corn issue that deserve answers.

Why did Heinz Wattie, Cedenco and Talley's not just pull out the crops themselves? The plants were still only seedlings and only 178 ha nationwide would be affected. Why did they not do the right thing?[13]

When the Government later gave approval for the companies to use the unplanted seeds from the Jubilee Lot NC9114 batch, did the companies go ahead and plant or sell seeds when they knew there was evidence of contamination?

Why were the business interests of the big companies given priority over the neighbouring farms, including sweet corn growers, who potentially suffered from the contaminated crops being left in?

Why was the organic sweet corn farmer not told? What economic value was placed on the risk of harming New Zealand's 'clean green' image?

And did the United States seed supplier, Novartis, accept responsibility and offer to compensate the New Zealand companies for its mistake in supplying contaminated seed?

This is the crucial issue of liability. When GE products cause financial damage to third parties, should it not be the one that introduces the products that is liable? If not, the risks will be quietly shifted onto other people and the environment. That is what happened here.

CHAPTER 3

GLITTERING PRIZES

Norrey Simmons – Novartis Seeds' PR representative – opened the annual conference of the Public Relations Institute in May 2000, just six months before she played a major part in the events in this book. Her speech to the conference, entitled 'Glittering Prizes', was about lobbying MMP governments.

She said that under MMP 'the system is more transparent, which is good, but it would be wise if people don't take too many minutes.' Talking about how PR people achieve their influence, she said 'We find officials are usually very co-operative people.... But we must never put them in a position where they will be compromised.' If people follow this advice, 'What can be achieved?', she asked. 'Glittering prizes!'[14]

Cabinet Minister Marian Hobbs, who later would be directed by Helen Clark to be 'lead Minister' on the GE corn contamination issue, had agreed to speak at the PR conference on the same panel as Simmons. Her subject: 'Are there tricks to reaching this Government?' It is hard to understand why politicians (or journalists) accept invitations to give tips to PR people. But they do.

Hobbs, besides reminding them that 'lobbying list MPs is very important', told the conference that the common experience of all the new Labour-Alliance Ministers was 'We're drowned in paper... we're short of time.' She told the audience of PR consultants that

Company PR representatives met government officials here, at the Ministry of Agriculture and Forestry headquarters in Wellington, at 5.30 p.m. on Wednesday, 29 November 2000. Norrey Simmons, working for the seed multinational Novartis, proposed that the Government accept 'US standards' for GE seed contamination.

the most effective way they could influence decisions was by supplying information: reports, helpful website addresses and so on. 'We don't want spin and opinion', she said, 'we are information hungry'.[15]

As it turned out, Simmons' and Hobbs' speeches spelt out pretty accurately the approach that would be taken soon afterwards by the company representatives in negotiations over the contaminated crops. They focussed their effort on the officials, who are conditioned to be very co-operative with business, and helped the Government keep the issue from the public. The busy Ministers, drowned in all their daily paperwork, accepted the facts and solutions that the PR people presented.

When Government eventually agreed to allow all the contaminated seeds to be used and harvested, and kept the incident quiet to avoid a backlash from the public, the companies had indeed achieved glittering prizes.

Norrey Simmons' PR company, Communications Trumps, is a Wellington-based company which she co-founded in 1987. Simmons had been in the news in 1999 because of her company's leading role in the genetically engineered salmon controversy. She has specialised in doing PR for a range of genetic engineering interests and was the most active of the PR people involved in the contaminated corn incident. In addition to individual clients (eg, Agresearch, the commercialised state research organisation undertaking GE cow research), she managed the pro-genetic engineering organisation Genepool, a supposedly independent GE education trust. Genepool had outlived its usefulness by 1999 and was closed down, following controversy over its links to Monsanto and using public science funds for promoting GE. Simmons was then involved in confidential meetings at the Wellington offices of the legal firm Russell McVeagh in which the successor lobby group, Life Sciences Network, was devised.[16] After its creation in May 2000, Life Sciences Network went on to have a huge influence during the Royal Commission on Genetic Engineering.

It is not known what part was played by the other two private PR companies involved in the contaminated crop issue. The second PR company was the New Zealand branch of the major United States PR company, Hill and Knowlton. The New Zealand executive chairman, Paul Hemsley, was the PR adviser to Heinz Wattie.

In New Zealand, Hill and Knowlton has a low profile. Overseas, the company has a reputation for its highly political public relations work. This is the company that 'managed communications' at the US Three Mile Island nuclear power station and presented bogus evidence to a Congressional meeting to get backing for the Gulf War (on behalf of its client the Kuwait Royal Family). Most recently, it has been hired 'to salvage Enron Corp.'.[17]

The third PR person involved was Cedric Allan, of Allan Boyle Associates. He was the PR adviser to Cedenco. At that same 2000 PR

Institute conference, Allan was voted in as the new president of the Institute, replacing Norrey Simmons. He had been outspoken in the Institute about the need for PR people to think more about ethical issues. Allan lives and works in Auckland and I do not know what his part was in the Wellington-based negotiations and lobbying over the contaminated crops.

In March 2001, shortly after the events detailed in this book, Allan gave a conference speech called 'ethical communications'. He asked his audience:

> *Do you need to actually be dishonest or to lie in order to be unethical? What about the ethics of withholding information about activities which your client or employer is undertaking? For instance, if you were working for ... a company which refused to recall a defective product? What would you do if such a situation arose and the client rejected your advice to remedy the problem and make a comprehensive public statement? Would you be justified in going public?...*
>
> *I suggest that it should be a growing part of the public relations function to be constantly advising our clients about the general ethical standards of the community in which they operate and the expectations of that community.*[18]

I wonder whether, in part, he was feeling unhappy about what he had seen happen over the contaminated crops.

CHAPTER 4

CRISIS MANAGEMENT

Environment and Biosecurity Minister, Marian Hobbs, was first informed about the seed contamination, verbally, on Monday, 13 November 2000. A few days passed while officials awaited the results from the retested Lot NC9114 samples and then the Government quickly moved into crisis management mode.

As soon as Prime Minister Helen Clark heard about the situation, she shifted overall control of the issue into her own Department of the Prime Minister and Cabinet (DPMC), overseen by departmental head Mark Prebble. A DPMC policy adviser, Ruth Wilkie, was given the job of hour by hour management of the issue. Clark knew and trusted Wilkie (they have mutual links through the family of lawyer-lobbyist Geoffrey Palmer, with whom Clark works closely) and Clark felt Wilkie was reliable and would be able to manage the sensitive issue.

A structure was established whereby the writing of policy advice and preparation of PR strategies would be co-ordinated within the Ministry for the Environment, with specialist advice provided by staff of the Environmental Risk Management Authority (ERMA) and the Ministry of Agriculture and Forestry's Biodiversity Authority. Advice from the Ministry of Foreign Affairs and Trade was also sought about World Trade Organisation restrictions and issues concerning the relationship with the United States Government.

Strict secrecy was imposed so that only a handful of staff in each agency even knew something was going on. Marian Hobbs was already

Mark Prebble, head of the Department of the Prime Minister and Cabinet — Helen Clark took control of the issue, moving day-to-day co-ordination of the Government's response into the DPMC, her own department. (Photo: Copyright The Dominion)

the Minister responsible for all three agencies, so Helen Clark appointed her 'lead Minister' and spokesperson for the Government on the issue. However, while the staff of these agencies would be doing much of the work and reporting to Hobbs, everyone understood that the decisions were being made, via the DPMC, by the Prime Minister. Ruth Wilkie would be the centre of the structure, overseeing all actions. As a relatively junior policy adviser, she was in no position to challenge what the Prime Minister directed. Her role was to manage the detail and be the conduit through which Helen Clark could be kept informed and issue instructions.

Helen Clark's initial judgement on the incident was decisive and sound. She told officials that, whatever the precise final policy decisions were, the 'bottom line' was that the crops must be pulled out. In the first stage of the management of the crisis, that is what the public servants involved expected would happen and planning began on how to achieve it.

Surprisingly, although there were complicated processes in place for considering applications for field trials of all manner of genetically

Action required

10. Irrespective of the medium term solution adopted, there is the issue of what to do with the existing shipment and the 178 hectares already planted out.

11. Bear in mind that in this case there are already several positive tests for contamination which can hardly be ignored. We are still giving thought to this and will discuss our thinking with the industry next week. However, one course of action may be to destroy the existing crop, but to allow batches of the remaining seed to be used **if** they pass the testing protocol referred to below.

12. In regard to the medium term, it is necessary for a policy approach to be agreed almost immediately. A second shipment of sweet corn seed is already en route to New Zealand and, in principle, the same issue arises with other types of seed. We are right in the middle of the planting season.

The head of of the Environmental Risk Management Authority, Bas Walker, wrote to Environment Minister Marian Hobbs saying that the several positive tests for contamination tests could 'hardly be ignored'.

engineered organisms – and many experiments had already been approved and were well underway around the country – no one in Government had got around to establishing the procedures if something went wrong. Under the Hazardous Substances and New Organisms Act (HSNO), there were provisions for appointing compliance officers with the power to order the destruction of unauthorised or risky trials and releases of new organisms. But the necessary regulations had never been put in place and no compliance officers had been appointed.

So, the first task tackled by Ministry for the Environment (MFE) staff was urgently drafting a Cabinet paper seeking formal approval for appointing compliance officers. An MFE memo to Ministers explained the action: 'MfE preparing late paper to approve HSNO compliance officers to Cabinet Monday 4 Dec (this is urgent so MAF and ERMA have the power to destroy the GM Maize if necessary).'[19] Cabinet duly approved it on 4 December, waiving the usual 28 day delay before regulations are gazetted, and a week later MAF's director of plant biosecurity, Richard Ivess, was formally designated as a compliance officer with the power to order the destruction of illegal crops.

The other urgent task co-ordinated by the Ministry for the Environment was working out the 'communications strategy'. At that stage the officials assumed that the Government would make a public announcement informing people about the contaminated crops. A date was set, only days away, for the official announcement. By that time they hoped to have compiled all the relevant information about the crops and what the government intended to do about them.

The idea of covering up the whole affair was not considered at that stage, partly because everyone involved assumed that such controversial news was bound to leak out to the news media. The main concern at that time was that the news would leak out before the Government had put in place the legal means to destroy the crops and prepared its own announcement. Work began on a 'communications contingency plan' covering who should say what if suddenly there were journalists phoning.

But the public servants were not going to be left to work out solutions and plans on their own. Once the follow-up tests had confirmed the presence of GE seeds in the batch, some of the companies involved, led by Novartis, the seed multinational, began urgent lobbying. Many details of the lobbying are not known, but the thrust of their efforts is clear.

Watch how, bit by bit, the Government moved to the position being advocated by the seed companies and their PR people. Within two weeks of negotiations beginning, Helen Clark's bottom line of pulling out the crops had disappeared completely. At that same time, the idea of being open with the public about what had happened was, in stages, delayed, then watered down and finally dropped.

On Friday morning, 24 November 2000, two weeks after news of the contaminated crops first reached the Government, a meeting was held between government officials and representatives of the main companies involved. The meeting participants were recorded as 'Novartis Seeds Pty Ltd. (Australia), Cedenco Foods Ltd., ARM (NZ) Ltd., Heinz Watties Australasia, Crop and Food Research, ERMA New Zealand and MAF.'[20]

ARM (NZ) Ltd may be Archer Risk Management, a company run by two former police detectives, which offers private investigation, crisis management and risk management to help clients 'protect their interests'. One of their clients is Talley's, so possibly they were representing Talley's at the meeting. Crop and Food is the Crown Research Institute that had tested the seeds for Cedenco and got the first positive GE result.

The companies' first request was that the issue remain secret at least until a course of action had been agreed. If or when it became public, they urged that the word 'maize' be used rather than 'sweet corn' in any government press releases, so as not to raise public concerns about food. They also urged that the negative-sounding word 'contamination' be avoided. A word like 'intermingling' was preferred.

The companies' main proposal at the meeting had much more serious ramifications. They argued that, internationally, contamination of seed stocks was getting more and more common. This meant the Government needed to adjust its laws to allow for the inevitable and unavoidable presence of some GE seeds mixed into future seed imports to New Zealand. Instead of the current regime based on the strict Hazardous Substances and New Organisms Act rules (prohibiting any unauthorised GE organisms being brought into New Zealand), they argued that the Government adopt 'US standards' relating to GE contamination.

The idea of 'US standards', which shortly after would form the basis of the public servants' advice to Ministers, was basically that the contamination should be accepted. The only practical New Zealand policy, it was argued, would be to allow a 'threshold' level (for instance, 0.1% or 1%) under which contamination would be tolerated.

This approach, which will be discussed in detail in the next chapter, has huge implications for New Zealand's natural environment and agriculture. Although presented as a practical and necessary policy, its effect would be to open a huge loophole in the regulation of GE organisms. The revised policy would open the borders to any new GE plants arriving accidentally as low level contamination of other approved crops.

It is very surprising that none of the agencies involved – all of which exist primarily to protect New Zealand's environment – did not argue more strongly against this approach. Essentially they were accepting that supposedly 'clean green' New Zealand, with its isolated and unique island ecosystems, should have no better protections and controls against GE contamination than the continental United States, a land where the use of genetically engineered crops in huge industrial farms is widespread.

Bas Walker, head of ERMA New Zealand, got the job of communicating the outcome of the initial negotiations with the companies to Marian Hobbs. His letter, written straight after meeting the industry representatives on that Friday, 24 November, put the case for a change of policy:

> *This series of events raises the fundamental policy issue, which is that if the* ***possibility*** *of contamination exists it is, in practical terms,* **not** *possible to verify with 100% certainty that it does not exist. No matter how extensive the sampling and testing, there will always be a residual margin of doubt* [his emphasis].[21]

He went on to explain that under the HSNO Act, 'the dilemma is that any level of contamination constitutes an unauthorised import for release.' His conclusion – which seems alarmist – was that 'If the standard is zero contamination then agriculture based on imported

GM Maize - Update - bullet points for POL

Overview

Some GM maize seed has been released in New Zealand. The contaminated seed was in a consignment that was certified GM free. The contaminated seed was detected in testing carried out by the importer who informed ERMA. Further plantings have been halted. A total of 178 hectares has been sown and it will be several weeks before it reaches the pollinating stage. Meanwhile officials and industry are working to determine the facts and consider the options for ensuring that the moratorium on GM organisms is not breached further.

Main actions under way:

1. Background information is being prepared by MfE, MAF and ERMA on:

   - Scope of problem (quantity of maize and other seed that could be GM contaminated).
   - Clarifying legal powers to act, liability and compensation issues
   - Policy options (zero tolerance and 1%)
   - Consequences (economic/trade/employment, environmental, health – note the main issue is public perceptions)

2. MfE, MAF, DPMC and ERMA meeting industry 1pm Wed 29 Dec. Note industry are being very cooperative.

3. MfE, MAF, DPMC and ERMA advising Minister late Wed on options/ consequences. May decide then that we have sufficient information and planning to go public on Thursday, or that more information is needed.

4. Communications strategy is being prepared - includes contingency and planned release scenario. Includes consultation with PM, key Ministers, and Coalition (and later on the Greens).

5. MfE preparing late paper to approve HSNO compliance officers to Cabinet Monday 4 Dec (this is urgent so MAF/ERMA have the power to destroy the GM Maize if necessary).

6. MfE is coordinating advice and will draft cabinet paper on broader policy issues as soon as feasible.

Ministry for the Environment

This Ministry for the Environment memo was written in late November 2000 while Helen Clark still wanted the crops pulled out. It shows planning for the crop removal and preparations for the public announcement of a decision to remove them.

seed might have to virtually close down, with huge economic implications.'

A lot of bureaucratic influence comes not from actually making the decisions, but from defining the options that the decision-makers consider. He laid out three possible solutions:

> *Option 1: Import seed only from areas (countries) which are known to be GM free.*
> *Option 2: Accept that absolute assurance is impracticable and therefore accept any shipments which pass very rigorous standards of testing.... Seeds meeting the standard would thus be "deemed" to be GM free.*
> *Option 3: Make an exception to the moratorium [which had stopped all commercial release of GE organisms] and allow a HSNO application for the release of common contaminants for particular types of seeds to be processed by the [ERMA] Authority.*

He dismissed Option 3, the proper process that his agency had been set up to implement, as too problematic. It would breach the moratorium on GE releases into the environment and 'open the door' to GM crops, he argued – both of which, of course, applied also to his preferred Option 2. He said importing seeds from GE-free areas, his Option 1, was a good idea but 'in many cases it may not be practicable.'

It was the familiar formula: offer three options, where option two, the middle way, is presented as a balanced compromise between problematic extremes. He recommended Option 2, saying that 'a decision would need to be made that a certain level (say 0.1% or 1%) of contamination was tolerable' and 'detailed operational procedures' established to implement it.

Walker noted that Ministry of Agriculture and Forestry staff had indicated they would probably support this approach as well. He said an urgent decision was needed because 'a second shipment of sweetcorn is already *en route* to New Zealand....We're right in the middle of the planting season.'

His letter also raised the subject of publicity. 'This is a very sensitive matter,' he wrote, 'and we would urge that no public statements are made until a course of action has been agreed. The industry are of this mind as well.'[22]

At this stage all the talk of thresholds was about future shipments of seeds, which still left the subject of the Lot NC9114 seeds. Walker wrote:

> *Irrespective of the medium term solution adopted, there is the issue of what to do with the existing shipment and the 178 hectares already planted out. Bear in mind that in this case, there are already several positive tests for contamination which can hardly be ignored.*

He suggested: 'One course of action may be to destroy the existing crop, but to allow batches of the remaining seed to be used if they pass the testing protocol. We are still giving thought to this and will discuss our thinking with the industry early next week.'[23].

'The industry' actually just meant the four New Zealand and overseas companies that were directly involved with the contaminated crops. Their interests (especially Novartis' international agenda of minimising controls on seed imports) were obviously not the same as the public interest nor the interests of other sectors of the agricultural and food industry. Yet they were being included in forming the advice going from the environment, biosecurity and risk management agencies to Government.

This is startling. It was not the environment and biosecurity agencies on one side and the companies, which were responsible for probably the biggest GE release in New Zealand history, on the other. It was the officials and company PR representatives working together. The PR people had constant access at the level of the Prime Minister's

Early kumara and corn gardens. Nga Kaihautu Tikanga Taiao, the Maori advisory body of the Environmental Risk Management Authority, advised in 1999 that any release of GE corn would risk transgenic pollen crossing with varieties of corn used by Maori for producing the traditional food kanga wai, kanga kopuwai and kanga pirau (fermented corn). (Alexander Turnbull Library, National Library of New Zealand, Te Puna Matauranga o Aotearoa. Ref No: G62651/1)

Department, where liaison was handled by Ruth Wilkie.

The next formal meeting between the company representatives and officials was a few days later on Wednesday, 29 November. The meeting was held at the Ministry of Agriculture and Forestry building, on The Terrace in Wellington, at 5.30 p.m. that evening. The main industry lobbyist at the meeting was Norrey Simmons, representing Novartis Seeds and taking the lead role for the businesses in the negotiations.

Simmons presented Novartis' position to the meeting. She stressed the need for secrecy, especially relating to the specific locations of fields containing the GE contaminated crops. She said they did not want the 'Wild Greens' to have an opportunity to go and pull up the crops themselves (a year earlier, in 1999, a group called 'Wild Greens' had pulled out a crop of genetically engineered potatoes in Christchurch). On the overall strategy, she said Novartis still wanted the New Zealand Government to accept the same kind of seed testing/threshold regime as operated in the United States.

The government officials and industry representatives talked through this option at length, then decided to get more information from Novartis in the United States before adopting a final position. The multinational seed company was having more and more influence over the policy development in New Zealand.

In the days following this meeting, the officials recommended, and the Government accepted, precisely the kind of 'tolerable' contamination levels that had been advocated. They agreed that the allowed GE seed contamination threshold in New Zealand would, under an interim regime, be set at 0.5% (that is, any shipments of seeds with under 0.5% GE contamination would be deemed 'GE-free' and allowed into the country). It was an immense shift in policy, yet it was made in an informal way without even referring the matter to Cabinet for a decision.

Simmons' other main point at the 29 November meeting was also very significant. The previous week, Novartis had informed officials that further testing had produced more evidence of GE contamination; and earlier the companies had said they were prepared to co-operate in removing the contaminated crops. But now Simmons argued that the test results were 'inconclusive'.[24]

Soon after, this small but crucial idea – that the tests were inconclusive – was adopted first by the officials, and then by the politicians. It became the justification for reversing all the Government's original plans for how to handle the contaminated sweet corn issue. If the GE test results were inconclusive, maybe there was no need to remove the crops after all. If the tests were inconclusive, maybe they need not say the contamination had happened. Three weeks had now passed without the story leaking out by itself, so if the results were inconclusive, maybe they could just keep quiet about the whole thing.

The commercial agenda of the companies coincided neatly with the pro-genetic engineering leanings of some of the senior officials and the political agenda of the Government, which was to avoid anything that might stir up controversy and public concern in the middle of the Royal Commission on Genetic Modification. It appears to have been the happy 'intermingling' of these different factors that led to the eventual decisions.

By 1 December, a month after the contamination was first discovered and also the day that public submissions to the Royal Commission closed, the new strategy was gaining momentum. Planning for the removal of the contaminated crops was quietly dropped. From then on all the bureaucratic effort went into selling the threshold idea to the Government.

As plans crystallised, the public and other affected farmers and crop growers were left completely out of the decision-making. But the big companies involved in the contamination were consulted regularly. For instance, on 30 November, the Ministry for the Environment sent the companies and their PR firms a draft contingency plan of what the Government would say if news of the contamination leaked out. Their views were sought on the draft government statement. The next day a final draft (reproduced at the back of the book) was sent to all the companies and their PR representatives, ensuring that if the news did reach the public the Government and the four companies representing industry would be speaking with one voice as they played down the incident.[25]

In the week of 4 December, while a series of organic producers organisations appeared before the Royal Commission in Wellington arguing how releases of GE crops could severely damage their industry, the officials finished writing a Cabinet paper to go to the following Monday's Cabinet meeting. The Cabinet paper was the end product of the rewriting of events that had begun a week before with Norrey Simmons arguing that the evidence of seed contamination was 'inconclusive'.

The public servants' advice to Ministers was now that, 'although initial testing suggested contamination at around the 0.5% level', after 'a more detailed evaluation' they now believed that 'if present at all, there is unlikely to be reliably discoverable GM contamination present in the particular [Lot NC9114] shipment.'

This was not correct. The whole of the coming chapter looks at details of the contamination to show that, as much as some people would have liked it to, it had not gone away.

CHAPTER 5

POLITICALLY MODIFIED SCIENCE

We are brought up to think of science as solid and incorruptible. In public debates, scientific facts are presented as being like rocks in the sea of conflicting opinions and beliefs. It is an attractive idea and it should be true. Disappointingly, it does not always work that way.

Producers of CFCs denied they were harming the earth's ozone layer long after the damage was evident. Scientists paid by oil companies and car manufacturers have repeatedly questioned the scientific evidence for global warming. Money can buy scientists and 'independent' scientific studies that have discovered smoking is harmless, that logging does not harm rainforests and that – another hugely lucrative business – genetic engineering is thoroughly controlled and safe.

It pays to remind ourselves over and over of the corruptibility of science (or, more precisely, scientific claims). For science can only help with important human questions when its insights are sensibly integrated with those from other spheres, notably ethics and the consideration of social responsibility.

It is very damaging to society when the public start to feel that they cannot trust people in positions of responsibility. Whether it is scientists, businesspeople, lawyers, public servants or politicians, we rely on these people to perform their roles in society ethically and responsibly. They are not just working for themselves or the people

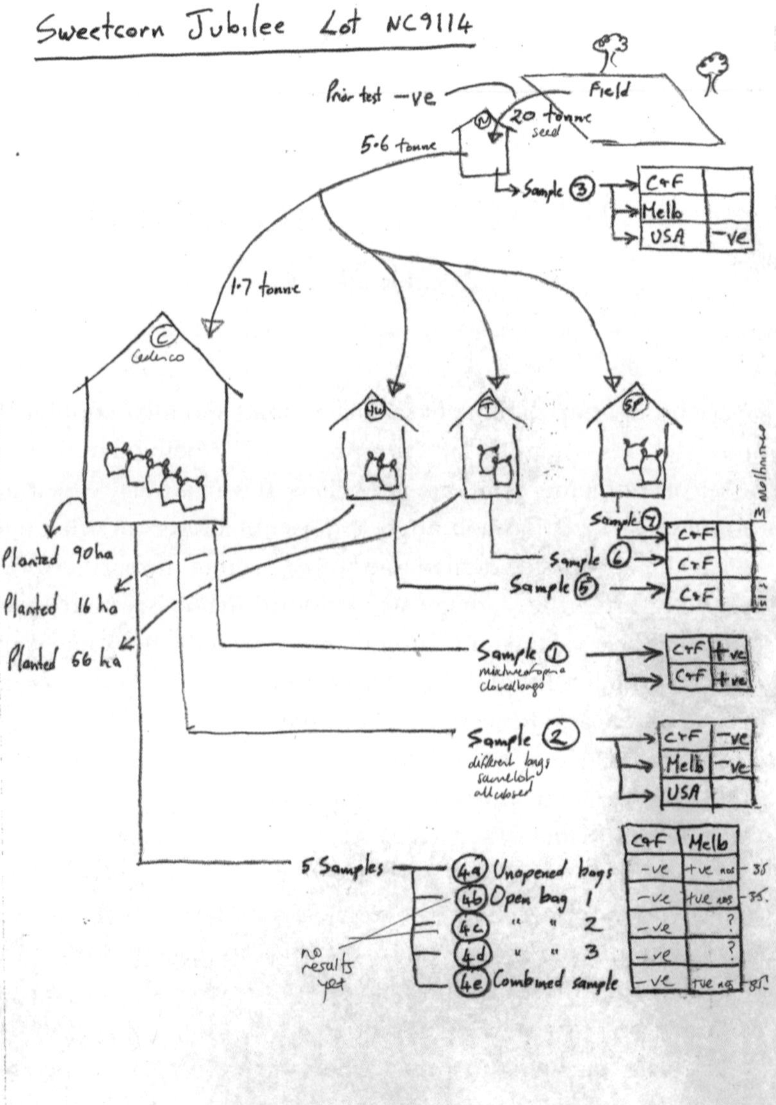

This hand-drawn diagram, by an official involved in the issue, records the various GE test results from different parts of the batch of Novartis seeds. With some tests positive and others negative, it was concluded that the GE seeds were probably spread unevenly through the batch.

who pay them; they are part of what makes democratic society possible. An important part of the cohesion of society is about the bond of trust (or lack of it) between ordinary people and people in these positions.

In the case of the contaminated corn crops, the businesspeople, PR consultants, public servants and politicians involved had one possible scientific conclusion that suited them and one that did not. The conclusion that suited them was, as we know, that once the scientists had checked more carefully, there turned out to have been no 'reliable evidence' of GM contamination in the Lot NC9114 sweet corn seeds after all.

This chapter picks through the evidence to explain why this was not true. Yet the people who were responsible to the public to make good decisions on these issues worked together to present the opposite 'fact' to the Cabinet and ensured that no one could contradict them, or provide different advice, by keeping the entire issue very secret. There were no hearings or submissions to test the evidence – as happens in all other cases relating to GE decisions – and so the convenient idea that the contaminated crops might not be contaminated was repeated between the officials until it felt true.

When the full 20 tonne batch of Lot NC9114 sweet corn seeds was tested by Novartis in the United States in August 2000, the test result came back negative. A few weeks later, when 5.6 tonnes of the batch arrived in New Zealand and 1.7 tonnes went to Cedenco Food Ltd, Cedenco sent samples to the Crown Research Institute Crop and Food laboratory at Lincoln for testing. Those were the first positive results. Later Cedenco retested its batch and again got positive results. Heinz Wattie, Talley's and Seed Distributors apparently also did tests of their batches, but their results are not known. The known results were summarised in a hand-drawn diagram, shown opposite.

On 24 November, by which time the follow-up test results were known, the head of ERMA, Bas Walker, advised Marian Hobbs to 'Bear in mind that in this case [Lot NC9114], there are already several

POLITICALLY CONTAMINATED SCIENCE

1 November 2000 - First positive GE test result of Cedenco seeds by Crop and Food laboratory.

24 November 2000 - Novartis, the US seed supplier, revealed that 'further extensive testing' had also produced positive results of GE contamination, *Report from industry meeting.*

24 November 2000 - 'Bear in mind that in this case, there are already several positive tests for contamination which can hardly be ignored.' *Bas Walker, advice to Marian Hobbs.*

29 November 2000 - 'Some GM maize seed has been released in New Zealand.' *Ministry for the Environment Cabinet committee briefing paper.*

30 November 2000 - 'In essence some of the tests have shown the presence of a genetic sequence which is strongly indicative of a genetic modification...' *Ministry for the Environment confidential question and answer sheet on the incident.*

5 December 2000 - 'With a considerable degree of confidence it can be concluded the Novartis sweet corn Lot NC9114 has less than 0.5% GM contamination and hence, judged by that standard, does not contain a new organism.' *Donald Hannah, ERMA Manager of Science and Research.*

8 December 2000 - 'If genetically modified material was present at all, it was unlikely to be more than 0.04% of this.' *Ministry for the Environment confidential question and answer sheet on the incident (later draft).*

positive tests for contamination which can hardly be ignored.'[26] A Ministry for the Environment briefing on the results said that 'In essence some of the tests have shown the presence of a genetic sequence which is strongly indicative of a genetic modification....'.[27]

At this stage of the handling of the issue, when officials knew Helen Clark wanted the contaminated crops destroyed, they were happy to say that the seeds were contaminated.

But then the politics shifted. Once the Government accepted the industry's proposition of a contamination threshold, the issue

8 December 2000 - 'There is no reliable evidence of GM contamination.... If present at all, there is unlikely to be reliably discoverable GM contamination present in the particular [Lot NC9114] shipment.' And, 'A detailed evaluation has suggested that there is negligible contamination.' *Cabinet Paper signed by Marian Hobbs.*

15 December 2000 - 'If genetically modified material was present at all in that shipment, the amount present was below the level that could be reliably detected.' *Ministry for the Environment confidential question and answer sheet on the incident (final draft).*

18 December 2000 - 'No evidence of GE crops growing in New Zealand.' *Key points for Minister.*

19 December 2000 - A 'detailed evaluation concluded that with a high degree of confidence that, if present at all, the GM material was at levels below that which can be reliably detected.' *Marian Hobbs' media statement.*

28 February 2001 - 'Tests could not confirm whether or not GM material was present.' *Government letter to Royal Commission.*

Note how the 'scientific' conclusions metamorphosed over a six week period. You can observe the exact point, at the end of November, when the politics changed and it became important to play down and then deny the results. It is a study in the vulnerability of scientific fact to politics.

was not whether or not there was contamination but whether it was below the newly adopted threshold and could be ignored (i.e. the seeds thus 'deemed' to be GE-free). Without any new testing, a review of the test results was conducted by ERMA, which became the basis of subsequent written advice from the other agencies and the Ministers' offices.

The terms of reference for the review, which was done by ERMA's Manager of Science and Research, Donald Hannah, were to decide whether the contamination level in the sweet corn seeds was above or below the new and arbitrary 0.5% level. He concluded: 'With a considerable degree of confidence it can be concluded the Novartis sweet corn Lot NC9114 has less than 0.5% GM contamination and hence, judged by that standard, does not contain a new organism.'[28]

This conclusion, combined with the 0.5% standard, had dealt neatly with the problem of the contaminated crops. It shows how flawed the threshold approach is. Judged by the new contamination standard, the crops were by definition 'GE-free' (even though they were not) and so they no longer had to be destroyed. Just add a drop of Policy Pragmatism and those difficult stains disappear!

But in case some Ministers, and later the public, felt uneasy about this sleight of hand, it became desirable to reinterpret the scientific findings. Dr Hannah's careful report was misinterpreted by other officials until they felt justified misleading Ministers by telling them that there turned out to be no reliable evidence of contamination of Lot NC9114 after all. The whole sorry progression of 'facts' about the contamination is set out in the box on pages 52 and 53.

According to Hannah's report, there were 12 separate samples taken of Lot NC9114 and sent for GE testing. Some of the results from the other companies had not been made available for his report. Of the data available to him, there was a mixture of positive and negative results. This is consistent with quite a low concentration of contamination. For instance, for three of the samples, the Genescan laboratory in Melbourne got positive results while the Crop and Food laboratory at Lincoln got negative results. It turns out that Genescan tested samples 100 times bigger than Crop and Food. Also, Hannah suggested, the contamination was probably not evenly distributed in the seed batch.

The test method used, called PCR, involves washing the sample to remove seed chemicals, grinding up the seeds and then extracting the DNA from sub-samples for testing. Each sub-sample has four PCR tests run on it so that the results can be compared. The PCR test kits can accurately detect GE contamination down to at least 0.1%

concentration. The report notes that it is much easier to say with certainty that there is some level of contamination in a set of samples than to say exactly what the level is in any one sample.

Dr Hannah made an estimate of the likely level of contamination by considering how many of the samples from the seed batch had 0.1% or more contamination (i.e. some level of contamination) and how many samples showed no contamination at the 0.1% level. He concluded that the overall level of contamination in the seed batch was about 0.04%. Contamination levels always sound small, but when spread across tens of millions of seeds (as in this case) it amounts to many thousands of GE plants scattered through the normal crops. Using his conservative estimate of 0.04% contamination, this amounted to about 15,000 GE sweet corn plants.

It is worth noting, in comparison, that the US Starlink contamination, which led to the US$100 million recall of corn-based products from shops and a sharp drop in US corn export prices, was estimated by the US Environmental Protection Agency (EPA) to involve a similar contamination level of 0.14%.[29] These EPA results were, by chance, released in the same week as the results of the testing of the New Zealand corn contamination.

Hannah cautiously considered the possibility that the positive results had been wrong (prefacing some comments with 'if there is contamination...'). He discussed the possibility that the test results were the result of lab error or dirty samples (giving 'false positives'). However, he found that results from unopened bags of seeds gave the same results as from opened (and so possibly dirty) bags. Similarly, he found that lab error was unlikely because the labs ran identical tests on 'controls' (seeds known to be GE-free) and these registered negative while the Lot NC9114 seeds registered positive. He also concluded that the likely contaminant was the genetically engineered 'Bt11' sweet corn.[30]

I have shown Hannah's report, containing the test findings, to genetic engineering specialists who say that the evidence of

EVALUATION OF INFORMATION RECEIVED FOR NOVARTIS JUBILEE SWEETCORN LOT NC9114

....Therefore, with a considerable degree of confidence it can be concluded that the Novartis sweet corn Lot NC9114 has less than 0.5% GM contamination and hence, judged by that standard, does not contain a new organism.

Donald Hannah
Manager, Science & Research
5 December 2000

This 'detailed valuation' of GE contamination results by ERMA scientist Donald Hannah was quoted throughout official papers. It looked at whether the GE contamination was below the Government's arbitrary 'acceptable' contamination level of 0.5%. Setting this arbitary level enabled the Government to deem the contaminated crops 'GE free'. This paper did not conclude, as later claimed by officials, that there had been no detectable contamination of the corn seed.

contamination appears conclusive. They say that even if there was any doubt, the obvious procedure was simply to rerun the tests until the results were undeniable.

This is also what one of the testing laboratories said. Discussing some test results that were 'at the limit of the system's detection', Genescan proposed that testing larger amounts of seed would allow definite conclusions about those particular samples. Hannah took up the point: 'Ideally in a testing regime repeated sampling should be undertaken until consistent results are achieved.... Multiple testing is desirable.'[31]

Yet, for some reason, later when people started claiming (incorrectly) that the results were uncertain, no follow up tests were ordered.

Hannah's instructions had been to assess whether the Lot NC9114 seeds complied with the brand new 'standard' of less than 0.5% GE contamination. He reached two main conclusions: one was that the overall contamination was below 0.5% and the other was that 'further sampling and testing is unlikely to change this conclusion'.[32] However, over the following weeks the conclusions were seriously misrepresented in crucial ways in the advice given to Ministers.

First, instead of saying that the 'detailed evaluation' had estimated 0.04% contamination, the Cabinet paper prepared by the Ministry for the Environment said there was 'unlikely to be reliably discoverable GM contamination present in the [Lot NC9114] shipment'.[33] Assuming that this was not deliberately misleading, the Ministry had completely misunderstood Hannah's report. Perhaps they thought that since the estimated 0.04% contamination was below the reliable detection level of PCR testing (0.1%), it meant that you could not say for sure that any contamination existed.

But Hannah's paper explained carefully that the 0.04% estimate was reached by combining results from different parts of the original seed shipment. Some of these samples were found to have contamination levels of 0.1% and above, and therefore demonstrated with reasonable certainty that there was GE contamination. But these results were put together with results from other parts of the shipment where no contamination was detected (thus Hannah's conclusion that the contaminated seed was unevenly distributed throughout the batch).

We can compare the New Zealand incident with the Austrian example, where GE contamination was found scattered through corn seed imports about six months after the New Zealand contamination. Although all the Austrian agencies confirmed with multiple testing that contamination was present, as in New Zealand it was unevenly spread through the batch. The Austrian Federal Office and Research Centre for Agriculture's Institute for Seed found that out of 230 samples that were taken, only 16 had contamination of 0.1% or more. They did not take this to mean 'not reliably detectable', or 'negligible', rather that the nature of contamination is that it is often present more in some parts of the seed batch than others. The Austrian authorities paid 530 farmers to pull out the crop.[34]

WHY THE TEST RESULTS WERE NOT 'INCONCLUSIVE'

When the Government justified leaving the GE plants in the ground by arguing that the sweet corn contamination results were inconclusive, they were politically modifying the science to suit their own ends. The tests were, in fact, conclusive. Here is why.

1. The levels of contamination detected were within the new standards of 'tolerable' contamination set by the Government and therefore, by definition, the sweet corn seeds were 'deemed GM free'.

This 'test' mixed policy and science. It is what the corn contamination Cabinet paper (Chapter 7) was referring to when it proposed the interim tolerable contamination standard and then said 'information on present sweet corn imports is that, against [the 0.5%] parameters, there is no reliable evidence for concern about GM contamination.'[35] In other words, there was reliable evidence of contamination in the 5.6 tonne seed batch, but overall it was below 0.5% concentration. Thus, elsewhere in the Cabinet paper, it was called 'negligible' contamination; not none or uncertain, but just below the arbitrary standard that had just been set. The 'negligible' level was still many thousands of GE corn plants – at concentrations which, in other countries, have led to the crops being destroyed and the seed batch recalled.

2. Officials argued that a 'more detailed evaluation' of the testing data showed that there was unlikely to be 'reliably discoverable' contamination in the sweet corn shipment.

'Reliably discoverable' refers to the technical capabilities of the testing equipment to reliably detect contamination. This argument was not denying that there were a series of positive GE contamination test results, but suggests that those results were unreliable because they were near the limits of what the test equipment could measure. This argument is factually incorrect. A 2002 MAF paper stated that 'PCR testing is so accurate that it can detect single genes but, in practice, the limit of detection is between 0.01% and 0.1% because of the limits on the size of the sample' – that is, 10-50 times more sensitive than the 0.5% threshold.[36] Testing laboratories say that the only limit on the system is sample sizes, which means testing is accurate to at least 0.1%. Thus the lab that used smaller samples when testing the corn shipment got less consistently positive results than the lab that used larger samples. Note that the new 'more detailed evaluation' did not mean they had done more tests, only that the existing results had been reinterpreted.

3. Officials argued that further testing would not give more certainty.

This argument was that the contamination, if there was any, was at such a low level that repeating the tests would give no more certainty. This is simply incorrect. The government scientist's report concluded that the contamination was below 0.5% concentration (which was all he had been asked to confirm) and he said that further testing was unlikely to change this conclusion (i.e. that it was under 0.5%). This conclusion was later misquoted in the Cabinet paper on seed contamination as saying that 'further sampling and testing is unlikely to change [the] conclusion' that there was no 'reliably detectable' contamination at all.[37] But if there had been doubts about the presence of contamination, which there weren't, the government scientist's report says that the normal procedure would be simply to rerun the tests. They did not rerun the tests.

The officials involved had no doubt that there was contamination in the sweet corn batch. Independent scientists who have reviewed the results for me have no doubt there was contamination in the sweet corn batch either. This is the advice Helen Clark was being given in the first weeks of the crisis. It was only after the 'tolerable' contamination level approach (suggested by the PR representatives of the four companies involved) was adopted that it became politically expedient to disguise the decision to tolerate contaminated crops in a fog of words about uncertainty. This is why science has to be conducted in the open, with results checkable and conclusions challengeable.

The Cabinet paper not only said that there was no 'reliably detectable' contamination, but that 'further sampling and testing is unlikely to change this conclusion'. Whether accidentally or deliberately, this also completely misrepresented the scientific analysis. Hannah had used those words, but when *he* used them he was saying that further sampling would not change his conclusion that the level was under 0.5%. He never said there was no 'reliably detectable' contamination – his analysis was based on there being a number of samples that had tested positive in ways that could not be dismissed as faulty testing. In fact, Hannah had made it clear that, if anyone had doubts about his conclusions, the way to clarify them was simply by doing some extra tests.

These shiftings of words made a nonsense of the science. A series of positive tests results showing GE contamination (which initially had been good enough to convince all concerned that there was problem to be dealt with) were suddenly no evidence at all. The obvious procedure of repeating the testing was declared to be of no use. The standard scientific procedure, if anyone really wanted more assurance of the contamination, was to rerun the tests – just as the Austrian authorities had done. But the New Zealand officials did not want more assurance.

And so Novartis' suggestion that the contamination results were 'inconclusive' became fact – and no further tests, which might have endangered this new fact, were deemed necessary. This was better than just 'deeming' the corn seeds 'GE-free' (i.e. under 0.5% contaminated), this was saying that contamination had never been conclusively established in the first place and thus made them GE-free.

As you would expect, of the 20-30 people who eventually came to know about the contaminated crops, not all were comfortable with the decisions and actions being taken. For instance, some of the departmental heads disagreed over how to handle the issue and some junior officials felt uneasy about covering it up (but did not feel in a position to stand up to the decision-makers).

One of the many things hidden during this period was that some of the people involved did openly object to the way the issues were being presented and to the decisions being made. Two in particular, Dr Lindie Nelson and Dr Oliver Sutherland, both members of the Environmental Risk Management Authority, felt very uncomfortable about what went on and put their opinions in writing. I have been given a copy of their confidential memo (reproduced at the back of the book).[38] Both reputable scientists, their memo (which was not written for politicians or the public, but purely for their colleagues on the Authority and its Maori advisory committee, Nga Kaihautu Tikanga Taiao) provides a very different view of the issue from the politically modified one we have seen so far.

Lindie Nelson specialises in resource management and agricultural economics and is a senior policy analyst in government. Oliver Sutherland is the Manager (Biodiversity and Ecosystem Management) at the government environment research organisation, Landcare Research, which has as its motto 'Making a difference for a truly clean, green New Zealand'. At that time he was deputy chair of the Environmental Risk Management Authority.

Their first concern was that the decisions about the contaminated crops had not followed the proper processes. All had been ad hoc decisions, coming down from the Prime Minister's office, whereas there were supposed to be strict procedures and processes for any new organisms. They began their memo:

> *Since our recent Board meeting (November 2000), major issues have arisen regarding the importation of GM-contaminated seed to New Zealand. These have led to various actions being taken by Government and by officials.... We have become increasingly concerned at the fact that some major decisions have been made (leading, amongst other things, to the planting of contaminated*

> *sweetcorn) which have had no active input from the Authority. It has prompted both of us to look more closely at the issues.... All of this increases our sense of unease at the Authority's lack of active input into policies as they have developed.*

It appears from the memo that most members of the Authority, which has a semi-judicial role regulating the introduction of new organisms, were not told about the sweet corn seed issue until after all the Government decisions had been made. They had been deliberately left out.

Nelson and Sutherland, writing in January 2001, summarised the situation:

> *In November of last year, a 5.6 tonne consignment of Novartis sweetcorn seed from the US arrived in New Zealand and was determined to have some contamination with GM seed. This has prompted two outcomes to date:*
>
> *(i) the Government has announced an interim proposal for a maximum acceptance level of 0.5% GM content in maize and sweetcorn seed...*
>
> *(ii) A range of tests (x10) was performed on the shipment of Novartis sweetcorn and an analysis of the results undertaken by Donald Hannah. He concluded that the seed... 'has less than 0.5% GM contamination...' and so was below the Government's maximum acceptance level of 0.5%. Accordingly approval was given (by officials/Minister) for the seed, which we understand had been planted at that stage, to remain in the soil.*

They noted pointedly: 'The plants are now maturing at known locations. It is still not too late for the plants to be destroyed before pollen is shed.'

Their memo then listed a series of concerns. The first was that

CONFIDENTIAL

concluded that the seed..... "has less than 0.5% GM contamination...." And so was below the Government's maximum acceptance level of 0.5%. Accordingly, approval was given (by officials/Minister) for the seed, which we understand had been planted at that stage, to remain in the soil. The plants are now maturing at known locations. It is still not too late for the plants to be destroyed before pollen is shed.

(i) **Maximum acceptance level of 0.5% contamination**

The interim approach ("standard") proposed is to establish sampling and testing procedures designed to detect with 99% confidence, that contamination is no greater than 0.5% (see Bas's memo of 21 December).[1]

The agreed approach is a de facto decision to allow release of grain with a maximum of 0.5% GM content. It is a very significant step. Our calculations suggest that for maize and sweet corn, this could translate to releasing up to 13.3 million GM plants per year throughout the NZ planting areas.[2]

We understand and support the practical need to define a standard, but are concerned about the apparent sidelining of the Authority and HSNO (the body and legislation set up to approve such releases) and the incongruity of strict risk evaluations that we undertake for field trial applications for contained growth of GM plants compared to 'back of the envelope' evaluations for this release. We are concerned at the apparent absence of any ex ante in-depth risk analysis to support the 0.5% acceptance level. Moreover, that this contamination level may be construed as being 'GE–free'.[3]

The moratorium is a voluntary one. No consideration seems to have been given to making an application to release GM contaminated seed, including the use of the Ministerial call in powers. Such an approach would allow the full HSNO evaluation process to be triggered.

The Authority has been very cautious in its approvals of field trial containment applications. We have agonised over the risks of very small pollen escapes and how we could prevent these. It is ironic to find that Cabinet and officials are taking a less cautious approach to a release decision. Further, given that PCR can reliably detect GM contamination down to 0.1%, why adopt the 0.5% level (there may be valid reasons but these are not apparent.)

Finally, the 'standard" could be used either as a "threshold" criterion or as a "yes/no" criterion. The documents provided so far suggest that intent is to apply the standard as a threshold.

Two members of the Environmental Risk Management Authority board wrote a confidential report strongly critical of the Government decision to allow 'a very substantial number of completely uncontrolled GM plants [to continue] growing in the field.'

the new 0.5% threshold for contamination opened the door to widespread genetic pollution in the New Zealand environment:

> *The agreed standard is a de facto decision to allow release of grain with a maximum of 0.5% GM content. It is a very significant step. Our calculations suggest that for maize and sweetcorn, this could translate to releasing up to 13.3 million GM plants per year throughout the NZ planting areas.*

Their next point was about the seemingly casual way that the environmental impacts of contaminated seeds had been dismissed:

> *We understand and support the practical need to define a standard, but are concerned about the apparent sidelining of the Authority and HSNO (the body and legislation set up to approve such releases) and the incongruity of strict risk evaluations that we undertake for field trial applications for contained growth of GM plants compared to 'back of the envelope' evaluations for this release. We are concerned at the apparent absence of any ex ante risk analysis to support the 0.5% acceptance level. Moreover, that this contamination level may be construed as being 'GE free'.*

They contrasted the 'back of the envelope' assessment of GE corn seed released uncontrolled into the environment with the Authority's approach to field trials:

> *The Authority has been very cautious in its approvals of field trial containment applications. We have agonized over the risks of very small pollen escapes and how we could prevent these. It is ironic to find that Cabinet and officials are taking a less cautious approach to a release decision.*

Quite aside from the question of environmental impacts, they questioned why a standard had been agreed to that permitted so much contamination. They could not understand why officials and the Government had chosen the 0.5% threshold, when it was possible to be much stricter.

'Given that PCR can reliably detect GM contamination down to 0.1%,' they asked, 'why adopt the 0.5% level? (There may be valid reasons but these are not apparent.)'. Then they challenged the whole rationale of the threshold system:

> *The documents provided so far suggest that intent is to apply the standard as a threshold criterion. That is, if contamination is detected, but is below 0.5%, it is acceptable. The alternative is to say that if the tests provide positive evidence of contamination, the seeds are unacceptable regardless of the level of contamination. In this alternative case, the release would be less because seeds known to be contaminated would not be planted. There does not seem to have been any discussion of the implications of the alternative application of the standard.*

They concluded by doing some calculations of the scale of the problem. They said that 'If we assume that 5 tonnes has been planted, then this is roughly equal to 35,000,000 seeds (there are c.7000 seeds/kg)' At the ERMA estimate of 0.04% contamination, they calculated this amounted to 14,000 GE plants. From their reading of the results, they felt that a more likely level was 0.08% contamination, meaning 28,000 plants. They noted that if contamination was near to the allowed 0.5% contamination level, the number of GE plants just from the one 5.6 tonne batch could be as many as 175,000.

> *Whatever the figure is, it is a very substantial number of completely uncontrolled GM plants now growing in the field. [NB. By comparison, the 0.4 hectare Pioneer transgenic maize*

> *field trial that ERMA approved in 1999 (but which wasn't planted) would have produced a total of just 1,450 GM plants – all pollen tassels bagged to prevent pollen escape together with a range of other controls.]*

What we see is that as soon as you look beyond the group of officials and PR advisers who had been playing down and defining away the problem, it was taken very seriously. This is not surprising. Seed imports are not some side issue in the efforts to protect New Zealand biosecurity. They are the most likely route by which genetically engineered plants could get into the New Zealand environment.

Nelson and Sutherland finished their confidential memo summarising their concerns as a set of 'observations'. They asked:

> • *Since we know that the Novartis sweetcorn is contaminated, is it acceptable that it should remain in the ground?*
> • *Should we, the Authority established to determine and assess environmental risks, be included in the decisions surrounding this seed/these plants?*
> • *Should we 'assert our authority' and step in at this stage, unsure, as we are, of who 'allowed' the Novartis plants to remain in the soil?*
> • *Is the standard being developed in a way that minimises contamination?*
> • *Should the public, who we know care about 'genetic pollution', be informed about the whole Novartis sweetcorn issue?*[39]

Why were these sorts of concerns and options not being raised by the officials or Ministers? In the next chapter we look at the bigger forces at work, which help explain why the Government did what it did and also show that there were other options available so that the crisis could have been handled quite differently.

Lindie Nelson and Oliver Sutherland's memo went nowhere. There was a special meeting of the Environmental Risk Management Authority in February 2001 to discuss the subject, but their paper had been placed last on the agenda, after a long series of presentations by some of the officials who had managed the issue and who tried to smooth the waters. The Authority expressed its dissatisfaction at the process, but could not change the decisions.

The corn plants, which by now were spreading their pollen, were staying in the ground. The new threshold regime was agreed to by Government.

In the tug-of-war between science and politics, politics had won.

CHAPTER 6

CONTEXT MATTERS

The greatest weakness of much news and public debate – and also secret government deliberations – is the way that issues are often presented and discussed without context. Issues are discussed without comparing a present situation with a past one, without comparing what's happening here with countries other than the US, UK and Australia, and often not even comparing a current statement with what the same person said previously.

This was certainly the case when the Government considered the contaminated seed issue. Some obvious questions were: how did the seed contamination issue in New Zealand compare with similar events in other countries? How had other governments responded? Is GE seed contamination really an inevitable fact of life? Was 0.5% contamination an acceptable threshold? Were there practical alternatives to accepting and denying the contamination that the Government could have adopted? In other words, were there more ways of looking at the problem and more ways of dealing with it than those proposed by the company lobbyists?

To understand the politics of the contaminated crops issue we need to see what happened in New Zealand as being just a small part of a much bigger international struggle over the use of genetically engineered crops. Essentially, the events in New Zealand were a little 'brush fire' in a world-wide fight by the multinational biotech corporations to impose commercial regimes optimal to their profitability. People like Norrey Simmons were just local hired guns paid to ensure that the conditions for maximum profitability prevailed in New Zealand.

First, putting Novartis in context. It was not just a seed supplier. According to trade publications, the company (which merged with AstraZeneca to become Sygenta on 13 November 2000) was 'the world's number one agro, chemical and seed company'. Together with Du Pont (owners of Pioneer Seeds), Monsanto, Aventis and Dow, it controls a large part of the world's GE and conventional crop seed supplies. It controls, on its own, about 10% of the world's plant biotech patents and is the third biggest producer of crop seed in the world. Its annual turnover is as large as New Zealand's total overseas trade.

Clicking your way around the company's website you see the operations and sales in 50 countries. The Australia-New Zealand website looks exactly the same as all the rest except that, lest we feel like just a line in a foreign corporation's balance sheet, where the other countries' websites say 'WELCOME', 'our' website alternately flashes 'G'DAY' and 'KIA ORA'.

World-wide commercial operations mean that the company has a huge financial interest in any government policies that might restrict its ability to sell its products. It is a straightforward commercial calculation that leads Novartis, and its sister GE corporations, to spend large sums through a variety of industry organisations and seemingly independent groups promoting genetic engineering and trying to discredit opponents of GE.

When we hear accusations thrown at New Zealand critics of genetic engineering – that they are anti-science, extreme, misinformed and even terrorists – such arguments have usually been imported from other countries where other industry funded 'life sciences' lobby groups are pushing corporate agendas.

For example, in 1999, *PR Week*, a public relations industry trade journal, reported that Novartis and other biotechnology corporations were gearing up for a new multi-million dollar campaign countering US public opposition to genetic engineering. Tony Minnichsoffer, Communications Manager for Novartis in Minneapolis, was quoted saying that the US food industry's 5000 trade associations 'need to work together as an industry' to broaden PR efforts on behalf of GE. Monsanto's preferred PR firm, Burson-Marsteller, warned that the GE debate in the United States 'is not in the crisis mode yet, but the potential is certainly there in this country.' Fleishman-Hilliard, a leading PR firm, said they expected their company alone to receive $2.5 million in contracts that year for PR work on 'crisis preparedness and issues relating to GE foods.' Fleishman-Hilliard recommended that food and biotech companies prepare for a major controversy over GE to erupt in the United States with a three-pronged approach: 'Anticipate potential issues; Drill with simulated situations to raise the crisis instinct within a company; Quickly deal with brush fires such as the [monarch] butterfly study.'[40]

When Norrey Simmons argued in Wellington for the adoption of 'US standards' on seed contamination, she was arguing for standards that would not cause the sorts of problems for the US seed corporations that, as we saw in Chapter 1, they were encountering more and more in other countries. That is, standards that accepted a certain level of contamination. When the New Zealand Government yielded to the lobbying and agreed to introduce 'acceptable' levels of contamination, we should be clear that the option accepted was the US seed companies' option.

We do not know what pressures or threats occurred at the Government level over the issue. It appears that the US Embassy in Wellington and Ministry of Foreign Affairs and Trade (MFAT) were both involved, raising issues about how New Zealand's World Trade Organisation (WTO) obligations affected decisions on the contaminated seeds. MFAT often argues within government for the 'rights' of foreign companies to operate in New Zealand without interference.

The WTO agenda on seed trade was seen on 17 April 2002 when the powerful Washington-based AgTrade Coalition lobby (which includes Novartis and many other companies) met with the new Bush Administration chair of WTO agricultural negotiations, Ambassador Stuart Harbison. Besides the usual call for further 'trade liberalisation' (that is, unrestrained access for these companies around the world), the AgTrade Coalition argued for keeping 'non-trade concerns, such as food safety and environmental issues, out of WTO agricultural negotiations.' (When our politicians explain how the WTO talks are all about helping countries like ours sell more lamb, it is worth recalling things like this.)

A specific item of the AgTrade Coalition agenda for the meeting was the 'removal of phytosanitory barriers pertaining to the grain and poultry industries', which, translated, means getting WTO rules to stop countries like New Zealand blocking imports of GE contaminated seeds for crops and poultry feed.[41] The American Seed Trade Association is quite open about this agenda. The *second* goal on its website is promoting 'better seed to produce better crops for a better way of life'. But its *first* goal is to be 'an effective voice for action in all matters concerning the...free movement of seed... throughout the world.'[42]

An international context also helps us to understand whether Novartis and the New Zealand government officials were right when they advised that seed contamination was now inevitable – and whether, as Lindie Nelson and Oliver Sutherland strongly disputed, the 0.5% threshold level for seed contamination was appropriate or even necessary.

The 8 December 2000 Cabinet paper on the seed issue invited Ministers to 'note' that 'inadvertent GM contamination (at low levels) in seed imports appears to be unavoidable, and increasingly likely to occur.' Recommending the setting of allowable contamination thresholds, it said that 'for purely practical reasons.... the present

A GE contamination level that is only part of a percent sounds low. But, over the whole crop, a conservative estimate is that over five tonnes of GE corn were produced. That's the equivalent of about 10,000 cans of genetically engineered corn (mixed into ordinary sweet corn) sold to New Zealanders and export markets. Note that by 2002, Heinz Wattie was assuring customers that all produce – including cans like these – was from GE-free sources.

interim proposal is for a maximum of 0.5% contamination in sweetcorn.'[43]

The paper says 'It needs to be clearly understood that, if this were done, New Zealand would be among the international leaders in setting up of systems for providing assurances that GM contamination in non-GM seed imports does not exceed defined low levels.'[44] This was perhaps intended to sound like New Zealand would be doing something admirable and environmentally responsible. But it means what it says, which is that the Government was being asked to be an international leader in adopting the US seed companies' preferred strategy of accepting 'low' level GE contamination as inevitable. New Zealand could then be cited as an example of low standards when other governments were being lobbied. Up until this time New Zealand policy had been that no GE contamination was permitted.[45] But corn seed shipments arriving the following year for the 2001-2002 growing season were accepted providing they had less than the 0.5% GE contamination.

Other countries were offering the leadership of a more environmentally responsible kind. While New Zealand Ministers were being led to believe that 0.5% was a necessary and practical low level for contamination, Austria already had much stronger regulations. In 2000 a maximum of 0.5% contamination was permitted *if* the GE contaminant was a GE crop already approved for growing (i.e. potentially already in the Austrian environment from before public opinion moved against GE crops). But it was 0% for any unapproved seeds (as the GE corn seeds were in New Zealand).

Since that time, Austria has moved further and, in 2001, passed a law to protect its conventional farmers from all GE contamination. Called 'Saatgut-Gentechnik-Vererdnung BGBlII478/2001', it sets a 0% GE contamination level for all seeds. Without exceptions. This means that, to the best ability of GE testing technology, no contaminated seeds are allowed into Austria.

In case this sounds like an isolated case, in January 2002, Italy's centre-right Government announced an equally strict policy. Farm Minister Giovanni Alemanno said 'The position that we have on seeds is a position of "zero tolerance" within the technical limits.' He went on to say that the Government would commit at least 50 million euros for customs and other measures to ensure seed imports were GE-free.[46]

In 2002, the European Union still had no binding directive on seed contamination, but an interim regulation allowed for 0.5% contamination of conventional seeds with GE seeds *if* the GE seeds were already approved for planting in Europe (i.e. even where an approved GE corn crop was growing next door to a conventional corn crop, the conventional crop still had to have less than 0.5% GE contamination). This is in effect a food contamination issue, relating to the eventual harvest. But for GE seeds that are not approved in Europe – i.e. when it is still an issue of keeping them out of the environment – the regulations stipulate that no GE contamination of conventional seeds is allowed at all – meaning a threshold of 0%. Looked at in this context, 'clean green' New Zealand could have done much better.

The New Zealand officials argued against zero contamination, saying that it was impossible to measure well enough to enforce such a policy. While literally true, this argument is disingenuous. Seeds can already be tested down to at least 0.1%, much less than the agreed 0.5%, and the test technology will continue to improve. New Zealand could easily have adopted a standard like Austria and Italy's: 'zero tolerance within the technical limits'. That does not guarantee that GE seeds cannot get in, but that is not the point. New Zealand biosecurity laws involve zero tolerance for poisonous spiders and snakes even though no border controls can guarantee they will not get in. It simply means the authorities aim to stop all they can at or before the border and then take measures to dispose of any unapproved species that do get through.

Under the interim 'standard' used for New Zealand's 2001-2002 growing season, if a shipment of corn had arrived registering 0.49% contamination, it would have been 'deemed' GE-free and allowed in for planting. During that season two shipments were refused for being above the 0.5% level: a consignment of corn for poultry feed and one small parcel of corn seeds (2.7 kg) from the United States. The obvious question to ask the Government is: how many seed consignments registering positive GE contamination were allowed into New Zealand and planted out under the 0.5% 'threshold' system that year?

As we will see in the final chapter, the entire threshold system was subsequently dropped – reinforcing the impression that it was only ever introduced to justify doing nothing about the Novartis Lot NC9114 contaminated corn crops.[47]

By 2002, Heinz Wattie and Talley's were publicly guaranteeing GE-free produce, as shown. But in the summer of 2000-1, the Government allowed the contaminated crops to keep growing and be harvested. The companies quietly sold them to an unsuspecting public.

When the December 2000 Cabinet paper quoted above said that 'for purely practical reasons...the present interim proposal is for a maximum of 0.5% contamination in sweetcorn', Ministers may have thought this was some sort of technical practicality, such as the limitations of the testing equipment. One paper prepared for Marian Hobbs stated that 0.5% was 'close to the limits of reliable detection'. However, the scientific paper prepared by the Environmental Risk Management Authority on GE testing was clear that 'based on advice received from the laboratories and other practitioners in the field,' the testing could detect 'well below the 0.5% level.'[48] Why, then, was 0.5% seen as the 'practical' level?

The answer may lie in one of the confidential papers prepared relating to the Lot NC9114 seeds. The 30 November 2000 paper talks about setting a 'practical level' of contamination, but explicitly states that this should be 'a practical level that the major seed producing nations [read, the United States] can meet'. It warns that if New Zealand were to set a tolerance of 0.1% there would be a high probability that the major seed producing nations would not be able to 'easily' meet it. It goes on: 'it may not be in the interests of major producing countries [multinational seed companies] to go to extra trouble and expense to meet our entry requirements.' Moreover, with strict border controls, 'there may be a backlash where New Zealand was no longer perceived as a convenient cost-effective place for other countries to bulk up seeds.'[49]

It could not be stated more baldly. The rationale for the 0.5% contamination control level was that it was the one with which the multinational seed companies could 'easily' comply. And being perceived as low-cost off-season growing space for the same companies was more important than environmental safety standards – never mind that other countries were seeing it as commercially beneficial to adopt strong policies for ensuring GE-free production.

This still leaves the question of 'inevitability' to consider. Are the companies and officials right when they say that New Zealand is too small to stop the tide of seed contamination? When we hear people talk of inevitabilities, it always pays to be suspicious. Inevitability is the great defence of the indefensible. Can New Zealand really not access uncontaminated seeds? Again, the question is much clearer once we look at the international context.

While seed contamination has been a growing problem overseas (and the multinational seed companies have a vested interest in turning what they *don't* see as a problem into inevitability), growing even faster has been the public rejection of genetic engineering in food. This trend has led to one major food company after another switching from supporting GE food to declaring their products GE-free and setting up produce tracking systems to prove it.

When you listen to genetic engineering lobby groups, they present a view of the world where there are realistic business people

Organic sweet corn grower, Gisborne. Around the world, more and more people are wanting GE-free food. This trend suggests GE contaminated seeds, far from being inevitable, could become commercially unviable.

on one side and well-intentioned but rather hysterical idealists on the other. So it is interesting to see that the hard-nosed New Zealand companies involved in the contaminated seeds incident – Heinz Wattie and Talley's – have since adopted strong GE-free policies for their products.

Cedenco, the Gisborne food processing company (now American owned), had, in 2000, already made undertakings to customers to provide GE-free produce. This is why it was doing the GE testing that first picked up the contamination in the seed batch. All these companies saw commercial advantage in GE-free policies and appear to have found ways of sourcing uncontaminated seeds.

This trend suggests that contaminated seed supplies, far from being inevitable, are increasingly not commercially viable. A perfect example of this is Novartis itself. Two months before the contaminated seeds arrived in New Zealand in October 2000, in a marvellous display of double standards, Novartis announced that it would no longer use any GE ingredients in any of the food brands it produced anywhere in the world. If the government officials had been looking at these trends, rather than focusing on the information and options from Novartis' lobbying, they would have seen options other than allowing the uncontrolled import of under 0.5% GE contaminated seeds.

Already there are states within the US, and countries elsewhere, positioning themselves as GE-free seed and crop producing zones, to meet the growing world demand for guaranteed GE-free products. What better place, one could ask, than islands in the south of the South Pacific for guaranteed GE-free seed production?

The seed contamination issue could have been seen as providing a market opportunity. Instead of leaving the negotiations having granted companies permission to plant out contaminated seeds around New Zealand (and to keep doing so under the business-friendly 0.5% import regime), officials could have left with proposals to investigate establishing a GE-free seed production industry. New Zealand already has a successful seed growing industry producing seeds, off season, for northern hemisphere crop production. Why not grow enough seeds for New Zealand, and then become a guaranteed GE-free seed source for other countries?

No doubt in the context of the secret meetings with Norrey Simmons and the other industry representatives, the government officials felt they were working out the best compromise for New Zealand, which demonstrates why secrecy in government is good for private but not public interest. If the decision-making had not been conducted entirely in secret, if other voices and options had been heard, it is likely there would have been different decisions made.

CHAPTER 7

MISLEADING THE MINISTERS

A month after the Prime Minister and selected Ministers first heard about the contaminated crops – a frantic 28 days of emergency meetings, negotiations and strict secrecy – the issue went to Cabinet on 11 December 2000. A few Ministers had heard varying amounts of the story by then, but for most this was the first time they had encountered it.

The Cabinet, which meets on the tenth floor of the Beehive Building, is supposedly where the big government decisions affecting the country are made. The Cabinet Manual says that Cabinet is 'the central decision making body of executive government' and is 'the ultimate arbiter of all government policy'.[50] Helen Clark, however, appears to have considered this issue too sensitive to be trusted to the Cabinet.

The real decisions on the contaminated seeds issue had effectively been made by a few senior Ministers and the officials before the Cabinet meeting. Instead, the Cabinet meeting became a complicated process of deception. Cabinet's stamp was being sought for decisions already made, but those in the know did not want to risk telling Ministers the whole story in case they did not agree or – even worse – did not agree with keeping it secret. What follows is a telling case study of how sensitive decisions can be manipulated.

For the first month after discovering the contaminated crops it had seemed inevitable that the news would leak out. But now the Prime Minister and officials had started to hope it could be kept quiet permanently. Knowledge of the issue had, up until then, been strictly limited to about 20 or so officials and a handful of key Ministers. Now the cover-up began in earnest.

Taking the issue to Cabinet, without really taking it there, was the first step in this process. Already, on instructions from Helen Clark, no hint of the crisis had been given to Ministers who might

The Labour-Alliance Cabinet: supposedly the 'ultimate arbiter of all government policy', Cabinet was actively misled about the contaminated crops and left out of the important decisions. (Photo: Copyright The Dominion)

not agree with keeping it secret. For instance, officials were specifically instructed to say nothing to Phillida Bunkle, an outspoken critic of genetic engineering, even though as Associate Minister for the Environment she should have been fully briefed. Now management of the Cabinet process began.

The Cabinet paper on the issue, signed by Marian Hobbs, was given to Ministers on Friday, 8 December, as part of the usual fat bundle of papers to read before the 10.30 a.m. Cabinet meeting on Monday morning. Except that the Cabinet paper was not actually on the main issue; that is, the thousands of GE sweet corn plants that would start to be harvested about five weeks later. The Cabinet paper that had been prepared for the Ministers was instead primarily about the general possibility of GE contaminated seed imports. Most of the content was about steps underway to make New Zealand an 'international leader' in setting up 'best practice guidelines' to deal with this problem by setting 'standards for tolerable levels of GM' (i.e. the 0.5% threshold).

Writing the Cabinet paper had been a joint process, involving a dozen or more people. It was contributed to by many of the people who had been working on different parts of the issue – with oversight by the Prime Minister's Department and the Ministry for the Environment. The evasions and various versions of the truth which they had helped develop over the previous weeks were thus brought together in the single document.

Once again, secrecy was the enemy of good government. When issues have been in the news, Ministers often get an opportunity to hear a range of views and information. On controversial issues, there will hopefully have been news about when the issue is going to Cabinet, debate about what the Government should do and journalistic efforts to find out anything the Government is not saying. Ministers know people will be watching their decisions. But on secret issues, the first time that most Ministers even hear about the subject may be when they are wading unsuspectingly through the week's pile of Cabinet papers, concentrating on their own issues. This was the case with the contaminated crops. It is not hard in such a situation to slip something through.

In this chapter, we work our way through the twists and turns of the Cabinet paper to see how approval was gained for something at which Cabinet may well have baulked, had the paper been more honest and open. Welcome to the world of Sir Humphrey.[51]

The Cabinet paper – you can read it in full at the back of the book – is a carefully crafted piece of work that achieves its purpose more by what it does not say than by what it does. The first striking act of 'not-saying' is the way in which it mentions the contaminated sweet corn seeds. As we will see, the paper says just enough so that, if ever challenged, the Government could say Cabinet was informed, when the subject was actually played down to a point where many Ministers probably didn't even notice it.

The second striking feature of the paper is that no options and no choices were presented to the Ministers. It is only after reading the document a couple of times that you realise the Ministers were not actually being asked to make a decision at all. They were only being asked to 'note' actions already underway. The sole point of the Cabinet paper appears to have been to persuade them to sign up to the decisions already made by the Prime Minister and Marian Hobbs to accept 'low level' GE seed contamination in New Zealand – and, by accepting those decisions, unknowingly to give their blessing to a decision to leave the contaminated crops in the ground.[52]

What would you expect a Cabinet paper to say on a controversial issue like this? I would have thought that right at the start Ministers would be told how much contaminated sweet corn was growing, where it was growing, when it was due to release pollen and that it was growing next to other sweet corn crops. I'd expect the Maori food implications to be discussed and the implications of GE contamination for New Zealand's reputation for clean produce. Then the paper might go through options, including pulling out the crops. But the Prime Minister had already made up her mind. None of those subjects was even mentioned in the paper.

The paper introduced the subject of the contaminated crops obliquely, by saying that 'A recent approach from several seed importers has alerted ERMA and MAF to the issue of contamination of seed with genetically modified material.' There was no mention of exactly what the seed companies were alerting ERMA and MAF about. A page later, Ministers were reassured that 'there is no reliable evidence of GM contamination' in New Zealand, and it was not until the third page that sweet corn was even mentioned.[53]

'The approach from seed importers', the paper says, 'appears to have arisen from a specific case of possible seed contamination.... involving a shipment of sweet corn from the US.' Appears to have arisen? *Appears* to have arisen? This was written by officials who had spent weeks frantically working on that specific case, expecting it to hit the news at any moment and preparing urgent plans for destroying the crops. Only a week earlier they had rushed through a regulation to empower a MAF biosecurity officer to order removal of the crops. Yet, now, they were playing down the whole incident as being just a possible explanation for why seed companies had responsibly thought to bring the general possibility of contaminated seed imports to the Government's notice at that time.

Having mentioned it, they immediately dismissed it by saying that although initial testing suggested 0.5% contamination, 'a more detailed evaluation concluded with a high degree of confidence that, if present at all, there is unlikely to be reliably discoverable GM contamination present in the particular shipment.'[54] We saw in the last chapter that this was not true and not what the 'more detailed evaluation' had said.

In other words, this was not a Cabinet paper about the contaminated crops at all. No information about them was given and no decisions were sought. The decisions that would legitimise leaving the crops in the ground were being presented as a pro-active response to a hypothetical problem of possible contaminated imports.

It was not until paragraph 19 on the fourth page of the paper that the crucial point was slipped in. Ministers were 'advised that, for purely practical reasons' an interim seed import regime was being introduced that permitted a maximum level of 0.5% contamination in sweet corn. Without explaining why only sweet corn was getting the interim regime, the paper later noted that:

> *Most sweet corn crops are already planted for the present growing season. The evidence suggests that for both these crops, and the small amount of seed that remains to be planted, if there is any GM material at all, it is not above the level in the proposal referred to in paragraph 19.*[55]

If a Minister had checked back to paragraph 19 and thought hard about it, it might have been clear what was going on. The cross-check reveals that by accepting the new 0.5% regime, the Government was allowing sweet corn crops 'already planted for the current growing season' to have up to 0.5% contamination. Not only that, they were also accepting that the companies could now go ahead and plant the 'small amount' of similarly contaminated seed 'that remains to be planted' (actually, over half of the contaminated Lot NC9114 seed batch). The recommendation on this at the end of the paper was

This Timaru seed distributer received one tonne of the contaminated sweet corn seeds. Initially the Government withheld biosecurity clearance, but later it gave approval for the contaminated seeds to be sold and planted.

'Did you know? Releasing a genetically modified organism in New Zealand without approval is illegal,' says this Ministry for the Environment education resource kit, showing corn growing in a well cared for environment. 'We have some of the strictest controls in the world.'

reassuring and non-specific. It invited Ministers to note that 'information on present sweet corn imports is that, against [the 0.5%] parameters, there is no reliable evidence for concern about GM contamination.'[56]

There is no evidence that any Ministers saw what was happening or objected to it. This is not surprising as the extracts quoted were buried within discussion about the future 'best practice guidelines' that the Cabinet paper was ostensibly about.

However, even with the real issue obscured, the officials still had some serious persuading to do. In the middle of the Royal Commission on Genetic Modification they were asking Ministers to 'note' (again, not 'agree') a whole new approach to 'how New Zealand defines "GM free"' that involved accepting GE contamination of seeds imports as 'unavoidable and increasingly likely to occur'.[57]

This part of the Cabinet paper is revealing about how the Ministers could be swayed. How do you persuade a government to do something it would not naturally do? Some of the standard arguments are: it will cost too much, other countries don't do it and business will be unhappy. Doubt can be cast on unhelpful evidence, while helpful claims can be uncritically accepted. Unwanted options can be presented as impractical, and desired ones as being the only possibility given inevitable factors. This is pretty much a summary of the Cabinet paper.

The most compelling arguments of course were about money and business. The Cabinet paper states, as a 'preliminary analysis', that the likely economic loss resulting from not importing seeds that might be contaminated could be $100 million per year.[58] The source of this very-big-number-that-cannot-be-ignored was not given. In fact, it was lifted from a paper written by the Ministry of Agriculture and Forestry (MAF) policy unit, during the preparation of the Cabinet paper, to justify introducing the allowable contamination thresholds.

The MAF paper looks serious and considered, but it is actually absurd. The $100 million per year is based on the assumption that the only way of eliminating the risk of contaminated seeds would be to ban all imports of seeds from GE crop producing countries. It jumps from here to the conclusion that New Zealand would have to stop importing seeds entirely (now from any country at all) and therefore whole areas of agriculture would stop. The $100 million is arrived at by adding together the total value of all New Zealand crops where GE versions exist overseas – all corn, maize, tomatoes, buttercup squash and canola – as if no further planting of these could go on if New Zealand was to stop importing contaminated seeds.

This paper is so analytically flawed, and typical of the many other 'it'll cost too much' reports used to sway government decisions, that I have reprinted it at the end of the book. Other officials had privately raised the option of sourcing seeds from countries or regions without

GE crops (and a stricter seed testing regime would have made a huge difference as well), but only MAF's alarmist economic loss predictions made it into the Cabinet paper.

The 'business will be unhappy' argument did not need to be spelt out. That was the point of the $100 million per year report and, one year into the Labour-Alliance Government, gaining the confidence of business was the foremost concern in most decisions.

Compared to the uncritical acceptance of the $100 million per year economic cost, the paper stated that 'there appear[ed] to be little possibility' of detrimental environmental consequences from that year's 'negligible' contamination. The issue of the environmental impact from allowing seed contamination in future years was not mentioned.

Appended to the Cabinet paper was a one-page 'overview' of the environmental and health risks of 'low levels' of GE contamination in sweet corn. The overview, written by ERMA, is reproduced in its full six paragraphs at the back of the book. Its main achievement, besides brevity, was appearing to make an assessment while studiously not mentioning the actual case of the Lot NC9114 contaminated crops. You might think an assessment of environmental impact would estimate the number of uncontrolled GE plants in the environment, discuss their locations and so the scale of the problem. It does not.

Instead, there are general conclusions such as how the risk of GE cross-pollination of conventional plants was very small because 'off-site cross-pollination...only occurs when there are other crops in the immediate (200 m) vicinity.'[59] No mention, though, that officials had known for weeks that some of the contaminated crops were located right next to conventional sweet corn crops, and in one case, next to an organic corn farm. Also no mention of the big 'potential long-term adverse effects'; GE issues – horizontal gene transfer, Maori traditional foods, scientific uncertainty and so on – that, as discussed in Chapter 2, ERMA had said would need to be faced if New Zealand moved from field trials to uncontrolled release.

After the Cabinet meeting, officials were told that the option of destroying the crops was now definitely off. Cabinet, without most Ministers knowing it, had accepted that the genetically engineered sweet corn would be left to grow and be harvested. (Photo: Photobank)

Almost every statement made by the Government on genetic engineering says its approach is 'precautionary'. Yet it was on the basis of these six paragraphs that the decision was made to allow the contaminated crops to pollinate and be harvested. The head of ERMA, Bas Walker, in an accompanying letter, at least conceded that the issue had 'not been subjected to our normal rigorous analysis and evaluation' and that this was merely an assessment 'at first blush'.[60] The rigorous assessments were done for contained field trials, and yet possibly the biggest uncontrolled release of genetically engineered organisms ever into the New Zealand environment had somehow been exempt.

But when the Ministers were presented with the issue, it was a very big $100 million economic loss on one side, versus only a 'possible', 'low', 'tolerable', 'very low', 'negligible' and 'unavoidable' level of contamination, 'if present at all', producing a 'very low' environmental risk, on the other. Which would the Ministers think was more important?

While referring repeatedly (and reassuringly) to 'very low' and 'tolerable' contamination, the paper did not explain what high or intolerable contamination would be. Hundreds of GE plants released? Thousands? Millions? There was no context. The argument was couched firmly in the advertising tradition of 'great buy...more economic gain for minimal contamination'.

To make strict contamination testing seem impractical, and 'some tolerance' of 'low level' contamination seem unavoidable, the Cabinet paper argued:

> *It is not possible to establish a regime which would provide absolute certainty that a batch of seed is GM-free. Absolute certainty would require every seed to be tested, and this would require every seed to be destroyed. Zero risk of the contamination of a shipment of seed with genetically modified seeds is therefore impossible...*[61]

This was obviously not written for scientists, but are politicians really that gullible? It is like saying that the only way to be 100% sure no poisonous spiders enter New Zealand is to search exhaustively through every incoming shipping container and strip search all arriving visitors. As with all other areas of border biosecurity, it is perfectly possible to devise practical testing procedures for keeping out GE seeds. Like other countries already do.

You have to question the professional ethics of public servants who would provide advice like this – and co-operate in hiding information from Cabinet. There is clearly a need for New Zealanders to demand a more independent public service, including reconsidering the changes that the 1980s Labour Government made to the public service, which undermined the concept of neutral public servants employed to provide free and frank advice.

While the Cabinet paper only 'noted' the issues outlined in this chapter, it did invite the Cabinet Ministers to make decisions on two subjects: agreeing to a briefing for the Royal Commission on 'the background and contents of this paper', and agreeing that Marian Hobbs make a public statement 'carefully outlining the dimensions of this issue'.[62]

In this context, the word 'careful' means the Ministers were being invited to go along with covering up the subject of GE contamination as much as possible to avoid controversy during the Royal Commission. In a section called 'Communication', but meaning the opposite, the Cabinet paper says cryptically that 'Because of the sensitivities around the Royal Commission and the voluntary moratorium in New Zealand, communication of the issue, and the steps underway to deal with it, will need to be carefully considered.'

It goes on: 'Both the public and the Royal Commission' will need to be 'informed of the full dimensions of the issue' in a 'clear and balanced way'.[63] We will see how 'clear and balanced' it was in the next chapter.

After the Cabinet meeting, the officials were instructed from the Prime Minister's Department that the option of destroying the crops (which the Ministers had not even been asked about) was now definitely off, since the contamination was under the new 0.5% threshold. They were also instructed that the 'announcement' would focus purely on introduction of the new seed testing regime. Word went around that Helen Clark had threatened Ministers that no leaks about the subject would be tolerated.

In the Cabinet discussion of the 'communication' strategy, it had been decided to delay the announcement about the new seed import regime until the week before Christmas, when Parliament would have finished for the year (to avoid questions) and the issue could most easily be buried. The Cabinet had been successfully managed and the final stage of the cover-up was ready.

CHAPTER 8

MEDIA MANIPULATION

There should be a rule in newsrooms, and especially in the Press Gallery where parliamentary journalists work, that any decisions announced by Government in the last week before Christmas will be scrutinised with the greatest suspicion. The rule would specify that extra attention be given to announcements that make the Government look responsible and good, since why would the Government be wanting these buried in the pre-Christmas rush?

Unfortunately there is no such rule in news organisations, only the pressures of a busy schedule of Christmas parties, darting out to do Christmas shopping and preparing for the holidays. That is why Governments have learned the simple and cynical trick of putting off announcements they do not want looked at too closely until that week.

The day chosen for the public announcement on the seed issue was six days before Christmas, on Tuesday 19 December 2000, by which time the thousands of GE corn plants were starting to mature and would soon be scattering their pollen. Marian Hobbs would be the spokesperson for the Government. A close look at the media management techniques used that day illustrates how, if you control all the information, an announcement can actually be a cover-up.

Earlier that year Hobbs had been interviewed by the *New Zealand Herald* about genetic engineering and spoke of her concerns about 'the whole dimension of the tangata whenua in relation to GE.' She said she was satisfied that the field-testing moratorium, running concurrent with the Royal Commission, would mean 'issues around safety and biodiversity are on hold'. 'What's not on hold,' she said, 'is ethics and the absolute importance of people being informed.'[64]

The preparation of the announcement was co-ordinated by the Ministry for the Environment and overseen by Ruth Wilkie in the Prime Minister's Department. A set of background papers was prepared for Hobbs, including two 'Question and Answer' sheets, the MAF $100 million economic impact report and information on past GE field trial approvals.

The Question and Answer sheet on sweet corn had gone through multiple drafts. An early draft had said 'In essence some of the tests have shown the presence of a genetic sequence which is strongly indicative of a genetic modification.' A week later it had been changed to 'If genetically modified material was present at all, it was unlikely to be more than 0.04% of this.' By the time it was given to Hobbs ten days after that, it misleadingly read 'If genetically modified material was present at all in that shipment, the amount present was below the level that could be reliably detected.'[65]

The day of the announcement was thoroughly stage-managed: who would say what, how could the announcement be made to look routine and how difficult issues could be ducked had all been carefully planned. No warning had been given to the media in advance. The first they knew of the issue, which had been the focus of frantic crisis management for the previous five weeks, was when Marian Hobbs' press secretary nonchalantly delivered a media statement around the Press Gallery at 9 a.m.

Making the announcement at 9 a.m. on a Tuesday morning, after most journalists had stayed late the evening before covering Monday Cabinet, ensured that there were not many people around the Press Gallery and some news organisations missed the story altogether. The journalists who were there were told that Hobbs would be

Environment Minister Marian Hobbs said she was satisfied that the GE field-testing moratorium, running concurrent with the Royal Commission, would mean 'issues around safety and biodiversity are on hold'. 'What's not on hold,' she said, 'is ethics and the absolute importance of people being informed.' (Photo: Copyright The Dominion)

available to answer questions in her office at 10.30 a.m.

The media statement was a distillation of all the 'careful and balanced' words that had been evolving over the previous weeks to obscure and hide the controversy. The Cabinet paper a week earlier had said that there was 'no reliable evidence for concern about GM contamination', and only someone who'd read very carefully might have noticed that this just meant contamination 'not above the level' to be 'tolerated'. The media statement, however, was the final step in totally denying that there had ever been any detectable contamination present.

The strategy of hiding the real story behind a dull-sounding import procedure announcement is immediately clear in the title of the media statement: 'New Safeguards Against GM Seed Introduction'. The 'good news' statement (which is reproduced at the end of the book) began:

> *New measures are being developed by the Government to guard against the inadvertent introduction of genetically-modified seed,*

> *the Minister for the Environment and Biosecurity, Marian Hobbs, announced today.*[66]

'We feel the need to strengthen border controls with respect to checking seed imports for possible GM presence,' Hobbs was quoted as saying in the statement.

In the fourth paragraph, she complimented seed importers for the 'very responsible approach' they were taking to 'managing the risks of possible GM content in batches of seeds'. This was a reference, we must assume, to the same food companies that were quietly leaving crops in the ground from the contaminated seed batch.

It was not until the fifth paragraph that the statement mentioned that these responsible importers had 'alerted the Environmental Risk Management Authority and the Ministry of Agriculture and Forestry to the possibility that some imported conventional sweetcorn seed might contain traces of genetically modified material.' And then, straight into the cover-up:

> *'Although initial testing suggested there might be minute traces of GM content, a more detailed evaluation concluded with a high degree of confidence that, if present at all, the GM material was at levels below that which can be reliably detected,' she [Marian Hobbs] said.*

The rest of the statement was some of the now familiar justifications for adopting a regime that accepted up to 0.5% contamination. 'Absolute certainty would require every seed to be tested' (and therefore destroyed). 'The only way to possibly eliminate the risk would be to ban all seed imports from all of the countries currently growing GM crops, a list which includes many of New Zealand's major trading partners.' 'Losses of an estimated $100 million a year [would] affect farmers and growers in many parts of the country'. The statement concluded by saying how 'cautious' and 'conservative' the Government's new approach was.

At 10 a.m., Hobbs had been given a final briefing from the senior officials who were handling the issue, to check that she had her lines off pat. The officials had written two scripts for her, both standard tools in highly controlled media strategies. One, known slightly disparagingly among communications people as a 'song sheet', was a deliberately short and simple set of points to guide Hobbs in what she should say. The other, called the 'back pocket notes', was a carefully structured guide on how to say as little as possible if she had to go beyond the song sheet.

Hon Marian Hobbs

Minister for the Environment and Biosecurity

Embargoed to 9am 19 December **Media Statement**

NEW SAFEGUARDS AGAINST GM SEED INTRODUCTION

New measures are being developed by the Government to guard against the inadvertent introduction of genetically-modified seed, the Minister for the Environment and Biosecurity, Marian Hobbs, announced today.

Officials are working on a seed testing regime and best practice guidelines to provide backup to the Hazardous Substances and New Organisms Act (which controls the deliberate introduction of new species) and the Biosecurity Act. They will also submit a special report to the Royal Commission on Genetic Modification.

"New Zealand imports seed from a number of countries where GM crops are grown," Marian Hobbs said. "Seeds which could be genetically modified include maize, sweet corn, squash, tomatoes, and canola. We feel the need to strengthen border controls with respect to checking seed imports for possible GM presence."

Ms Hobbs said a number of seed importers are taking a very responsible approach to managing the risks of possible GM content in batches of seeds. They

The media strategy was to hide the real story behind a dull-sounding import procedure announcement. Thus the 'announcement' was actually part of the coverup. It illustrates how, when you control the information, an announcement can actually be a coverup.

MARIAN HOBBS ON RADIO NEW ZEALAND RURAL NEWS 12. 30 P.M. 19 DECEMBER 2000

News Reader: Vegetable growers are among those relieved that the government's introduced a new testing regime for seed imports to prevent the inadvertent introduction of genetically modified seeds. The Environment and Biosecurity Minister Marian Hobbs says New Zealand imports seeds from a number of countries which use gene technology in some of their seed production, and while this country only imports seeds certified as GE-free, the Minister says there are no tests in place to check the certification. Seed shipments will now be checked on a random basis to, ah, hopefully stop any shipments that contain GE seed. The Vegetable and Potato Growers Federation says the industry is highly dependent on imported seed, so a sampling and testing regime is essential. It says the government's taking similar steps to many of New Zealand's trading partners, including Europe and Japan. And the minister's acknowledged that the governments move to vet seed imports for GM content is in response to concerns raised by its country's own seed industry.

Marian Hobbs: This has actually come from the industry. They are as concerned as anybody else that we are GE -free. We do a tremendous amount of testing... you call it cultivar development of new seeds in New Zealand, and the value to us is because we are GE-free. This is really important for the industry

This is not just because of the moratorium or those issues, it's really important for this industry to be so. So they are as concerned as anybody else to be so, because we want to be able to export our seeds.

News Reader: And Marian Hobbs says it would be impossible to guarantee every seed that enters the country is GE-free, but she says the testing regime will go as far as it can.

Marian Hobbs: You can never say that anything is totally GE-free, unless you were to test every seed in the consignment and, therefore, nothing to plant, because when you test the seed, you destroy it (laughs). So, we set the limits, ah, in one particular seed, and it's only an interim one, of testing, to a limit of 0.5%. And 0.5% is as tight as we can go to reliably say that there is neither GE seed present in the consignment.

Marian Hobbs on Radio New Zealand News. She used all the arguments on the 'song sheet' to justify why the Government had accepted 0.5% GE contamination. She said the government policy had 'actually come from the industry. They are as concerned as anybody else that we are GE-free.'

These PR documents, which together with the press release are reproduced at the back of the book, are illuminating about the way the news media (and Ministers) can be managed. The details of the Government's media strategy had been co-ordinated with the companies' PR people so that, if the story could not be contained to the basic announcement, everyone involved would be saying, and not saying, the same things. This Government-business collaboration included lining up at least one sympathetic industry group to chime in supporting the Government's announcement as soon as it came out. Nothing was being left to chance.

The song sheet, tactfully titled 'Key points for the Minister', was a mixture of real information, PR spin and, well, not real information. For example, point one was: 'No evidence of GM crops in New Zealand.' All the other favourite lines were there: the $100 million cost of not implementing the 0.5% regime, 'interim proposals are near limits of reliable detection' and so on. The song sheet even said, 'NZ not GM free...GM research for last 12 years; food in supermarkets has GM content', presumably, when Hobbs announced the acceptance of GE contamination, in order to play down the significance of the fact that New Zealand had never before (officially) allowed GE organisms to be released into the environment.

The back pocket notes were evasions in case a journalist started asking probing questions. To a series of possible questions, there were one or more answers for Hobbs, depending on how hard she was pushed or whether the journalist had some information from elsewhere. For instance:

> *Where are these crops growing?*
> *1. In a number of regions of New Zealand. I suggest you talk to the industry spokesperson, Peter Silcock.*
> *2. There are plantings in East Coast, Hawkes Bay and Marlborough.*
> *3. I do not know the actual sites where these crops are growing.*

The purpose of these options was, clearly, so the Government could release as little information as could be got away with. Contrary to what Hobbs had promised, it seems 'ethics and the absolute importance of people being informed' had been put on hold.

The media plans all worked perfectly. It appears from the media coverage that hardly any journalists bothered to go up to Hobbs' office. Most newspapers used a short story prepared by the New Zealand Press Association, which was almost word for word a slightly shortened version of the media statement: the PR staff's words were transferred almost verbatim onto the nation's newspaper pages. One paper tried slightly harder, inserting three paragraphs from a Green Party press release (which had responded to the Government one) into its own barely altered version of the media statement.

The newspapers, by simply transferring the Government media statement into their stories, at least included the paragraphs that hinted there had been some sort of initial concern about possible contamination. But the radio news coverage bought the 'good news' story completely.

The radio coverage featured Hobbs ad libbing on tape, not just the written 'quotes' from the media statement that were used in the newspaper stories. But there was no need for her to use the back pocket notes, because no probing questions were asked. The transcripts of one of the radio news items is in the box on page 96.

Hobbs followed her lines well. The song sheet said 'testing every seed means grinding up and destroying ability to grow'. Hobbs said on Radio New Zealand's midday rural news: 'you can never say anything is totally GE free, unless you were to test every seed and, therefore, nothing to plant, because when you test the seed you destroy it [laughs]'. The song sheet said 'Interim proposals are near limits of reliable detection'. Hobbs said 'It's only an interim, of testing, to a limit of 0.5%. And 0.5% is as tight as we can go to reliably say that there is neither GE seed present in the consignment.' She made the obligatory friendly comment about business (about the companies growing the contaminated crops) without needing notes: 'This has actually come from the industry. They are as concerned as anybody else that we are GE free.' She ended by going beyond her notes, leaving any doubts about the contamination behind: 'We do a tremendous amount of testing...you call it cultivar development of new seeds in New Zealand, and the value to us *is* because we are GE-free.'[67]

'Drowned in paper... short of time', Hobbs was letting herself be part of the media manipulation planned elsewhere: fronting for decisions made by Helen Clark and using arguments scripted by officials. Ultimately, it was Novartis' lines she was speaking. This reminds us that Cabinet Ministers, those 'ultimate arbiters of Government policy', spend a lot of their time reading speeches, signing letters and approving policy papers prepared by others. Unless they keep themselves well-informed by getting information from a variety of sources, and are determined on an issue, they are a bit like figureheads on the front of the ship of state.

The Government announcement was definitely not TV news material – too boring and complex. Any media statement where the second sentence begins 'Officials are working on a seed testing regime and best practice guidelines' is most likely destined for the bin. So, as far as I know, the story was not covered by television news. The back pocket notes covering the question 'Can we get film of these crops?' – 'I suggest you talk to the industry spokesperson' were not required.

The final stage of the contaminated seeds media strategy was what is called third party endorsements. This is the well-worn PR tactic where, if you think that the public may not wholly believe what you say, you get someone who sounds more independent to say it for you. It's remarkable how unsuspecting some news organisations seem to be when these independent voices pop up at the right time.

Officials had contacted and briefed the Vegetable and Potato Growers Federation in advance so that, shortly after the release of the Government's media statement, the Federation responded praising the Government's decision. The midday radio news item that interviewed Hobbs began: 'Vegetable growers are among those relieved that the Government has introduced a new testing regime for seed imports to prevent the inadvertent introduction of genetically modified seeds.' Not bad coverage for a decision that had actually been formulated to 'tolerate' the inadvertent introduction of GE seeds.

The media strategy had also included a special briefing for the Green Party's Jeanette Fitzsimons (illustrating the less than altruistic side of 'consultation'). Although suspicious about genetic engineering, Fitzsimons believed what she was told about the limitations and inconclusiveness of the corn testing and so gave support to the Government's move. Since she'd been notified in advance, she had a media statement – 'GE standards for seed protects New Zealand's market advantage' – ready to go out on the morning of the announcement. It was issued by the Green Party media unit before Marian Hobbs' press conference was finished.

The media statement 'welcomed the Government's move today to screen all imported seeds which are to be grown in New Zealand'. It said:

> *Possible contamination came to light when a company supplying an order of GE-free corn for export tested its imported seed to check it could guarantee its purity...*
>
> *'The fact that New Zealand was chosen as the place to grow this crop underlines the benefits we have always claimed of maintaining our GE-free reputation,' said Ms Fitzsimons. 'However today's developments show just how easily we could lose this advantage through inadvertant GE contamination. We must be extremely vigilant to preserve our rare GE-free status which is already bringing economic benefits'...*[68]

The Green Party co-leader Rod Donald, quoted in the *Press*, said 'the more stringent regulations were good. "In our view New Zealand's agricultural future depends on us remaining GM-free." He said the Green Party accepted that zero tolerance at the point of importation was impossible. "What is crucial is that if the crop is found to contain GM material, it is destroyed."'[69]

Both of the Green Party leaders had been successfully misled by the misinformation and clearly neither of them had any inkling there was a cover-up. They had been cleverly used as part of the Government's media plan.

The media strategy, if it had been written down, would have said something like this:

- Spin the announcement into a good news story about the Government acting to keep New Zealand GE-free
- Present the announcement as routine and unexceptional
- Mention the sweet corn contamination in a brief and inconspicuous way, concluding that no contamination occurred, to avoid any future claims of the Government not being open about it
- Do not mention the companies, locations or health and environmental issues
- Give the publicity a 'Government and business working well together spin', and emphasise how responsible and co-operative the seed companies have been
- Ensure that the story is dead and forgotten in 24 hours.

It worked precisely to plan. At the end of that day, there was a sense of euphoria and relief among all the people who had worked over the preceding weeks to control the crisis. The media had uncritically swallowed the announcement. Nothing had leaked out. No one had stuffed up. Even the Green Party had believed the stories.

They had got away with it.

CHAPTER 9

'BUSINESS-FRIENDLY' GOVERNMENT

I hate to say it, but this is vintage Clark. I've seen her on so many occasions go in with a firm bottom line, then the pressure comes on from business and she backs down.
– Senior government source on the contaminated crops, June 2002.

On 9 October 2000, Helen Clark gave a speech at Ellerslie Racecourse to people involved in New Zealand's food industry. The occasion was the annual Food Excellence Awards. She told the audience she had opened a new development at the Heinz Wattie factory in Hastings three days earlier and then spoke of the importance of environmental standards for the country's economic future.

'Environmental responsibility is important to New Zealanders and our acceptance of it has spin-offs for the food industry,' she said. 'We are seen as clean and green. People around the world believe that we deliver fresh natural products. There is room to turn this care for the environment into even greater commercial returns.'

> *We are also known as a country for our honesty, openness and integrity. Other countries trust our regulatory institutions and look with envy at our lack of corruption and open information systems.*
>
> *The time is coming to turn this into even greater commercial*

> *benefit for industries such as food, where consumers are increasingly concerned about food safety and are prepared to pay more for it and for quality.*[70]

This, in a nutshell, is what the seed contamination issue is about. Why should New Zealanders trust government and company assurances about an unpredictable new technology like genetic engineering if, when something goes wrong, it's cleverly hushed up? Why should citizens of other countries trust our 'clean and green' agricultural produce, if the Government would not remove a relatively small area of contaminated crop to protect the integrity of the rest?

Honesty, openness and integrity are not usually tested when everything is going smoothly. They are usually needed on precisely those occasions when it would be easier, more 'pragmatic', not to do the right thing.

It was only four weeks after this food award speech that Helen Clark was first told about the contaminated corn crops. Her first reaction, as we saw, was to do the right thing: have the contaminated crops removed and publicise what had happened. Then pressure came on from business.

Backed by senior officials, she opted to leave the crops in the ground and keep quiet about the incident: breaking faith with the public, local Maori people and neighbouring farmers – not to mention the food industry people whose futures depend on how 'people around the world' trust New Zealand's green reputation. At least 15,000 GE corn plants were harvested and sold to unsuspecting people. Helen Clark agreed to a new (and probably unlawful) import regime for the following season that legitimised GE contaminated seeds (by redefining them as GE-free) and even allowed the rest of the known contaminated seed batch to be used. Why?

Helen Clark was elected to government with the potential to be one of New Zealand's great Prime Ministers. In terms of caring about

Four weeks before being told about the contaminated corn crops, Helen Clark told a food industry meeting: 'We are seen as clean and green.... We are also known as a country for our honesty, openness and integrity. Other countries trust our regulatory institutions.... The time is coming to turn this into even greater commercial benefit for industries such as food.' This, in a nutshell, is what the contaminated crop issue is about. Why should people trust decisions about genetic engineering if, when something goes wrong, it is cleverly hushed up? (Photo: Copyright The Dominion)

issues and ability to lead, she is head and shoulders above Prime Ministers of recent history. Yet her handling of the seed issue appears cynical and dishonest. What is going on? My view is that two factors are at work, one large and tied up with international politics, the other about Clark herself and the style of New Zealand politics.

The contaminated seed issue arose at the end of the Labour-Alliance coalition's first year in government. After starting confidently, the Government had faced a concerted political campaign from business interests opposed, among other things, to its election promise of changing the Employment Contracts Act. Then came the 2000 'winter of discontent', the so-called 'crisis' when business confidence polls dropped dramatically.

Month after month the news media ran stories about the business confidence crisis, and succeeded in creating a sense that something very bad was happening. One news story titled 'Wiping smiles off faces' began 'After riding high in the opinion polls, the honeymoon is over for the Labour-Alliance coalition'. It spoke of 'gloom', 'pessimism' and a 'chorus of disapproval' from business.[71] The situation smelt of orchestration, and at the very least was enthusiastically encouraged by opponents of the Government in the media and elsewhere.

Looking back, the 'crisis' seems a bit ridiculous. The worst that happened was a fall in the New Zealand exchange rate (the value of the 'kiwi dollar' compared to other currencies), which caused a boom in exports and growth in the economy. But at the time the Government, and particularly Clark, became more and more preoccupied by all the bad publicity.

A senior government source, someone working close to Clark at the time, says that she was 'seriously rattled' by the idea that business opposition could destabilise her government. 'She was suddenly much more cautious about everything, always talking about how we mustn't get offside with business....I think she believed that after spending 20 years getting to where she was, with a chance to change things the way she wanted, she was at risk of losing it all.'

The Government began an urgent campaign to persuade business leaders that its social agenda was modest and that it was fundamentally a business-friendly Government. This campaign was epitomised by a Government-organised Business Forum at the end of October 2000. The forum, held at the Auckland offices of the legal firm Buddle Findlay, brought together a hand-picked group of major business leaders with Helen Clark and 10 other Cabinet Ministers.

Helen Clark's speech assured the audience that on most economic 'fundamentals' her Government had the same policies as the previous Government, then spoke about how 'much stronger partnerships are required between government and business':

> *My invitation to business today is to join us in that partnership...Business leadership is needed alongside government leadership. Government in the 21st century can lead, facilitate, co-ordinate, act as a broker and, where appropriate, fund and provide. But it takes two to tango.*[72]

You cannot blame the Government for wanting to do something to reduce the sense of conflict with business lobby groups. But these conciliatory words make light of a very serious issue, which is the growth of business influence over elected governments in democratic societies around the world. Business leadership alongside government leadership is like those countries with military leadership alongside government leadership. It is not how democratic societies are supposed to work.

However, from that time on, acceptability to business (meaning big business, not plumbers, restaurants and other small businesses) became a dominant consideration in all the Clark Government's decision making. She could act strong, even angry, standing up to the unreasonable demands of, say, environmental groups, but would bend over backwards to listen to the views of business leaders about what they felt was best for the country.

This was the context in which Helen Clark and her advisers received news of the contaminated crops, literally just days after that Business Forum speech. Things were just starting to look up after months of accusations of being anti-business. It was very important to Clark that her Government be seen to be business-friendly by the companies and country (the United States) involved in the contamination issue.

The meetings between government officials and the company representatives (with no other interests represented) are an excellent example of a partnership between government and business. Together they worked out the solutions most acceptable to the companies and even joint PR strategies. It is not surprising that, when business leadership joined Government leadership in the decision-making process, Clark ended up backing the options most favourable to the multinational agro-tech company.

The other contributing factor concerns Helen Clark herself. The issue is one of political leadership versus political management. Clark can be a very effective leader. When she leads, she is formidable. But especially after the business confidence crisis, the manager in her increasingly prevailed.

Quoting the senior government source again:

> *Clark can look strong when she wants to, like when she attacks someone, so people think she's fierce and resolute.... But, actually, her main problem is that she is too fearful. The policies of the last three years are littered with the results of fear.... You start making compromises, and before you know it you've made more compromises than you ever thought you would.*

The tools of a political leader include openness, confronting issues and taking the public with them. Great leaders do not just survive politically, they inspire others to go with them as they do what they believe in. The tools of a political manager are different: secrecy, avoiding trouble by compromising with powerful interests and using PR to handle the public. Political management was the defining feature of past National Governments. Stripped down to basics, it means deferring to power – to big business in national affairs and to the United States in international affairs.

On the contaminated corn issue, the leader in Clark wanted to be decisive and do the right thing, in part to preserve the integrity of the Royal Commission process. But the cautious political manager went for what Novartis wanted and what was politically expedient in the middle of the Royal Commission: quietly controlling the issue through her Department and leaving it to Marian Hobbs to front it in public.

The dominance of the political manager over the political leader is an important part of understanding the Clark Government's actions on many other subjects. This was the tragedy of David Lange's time as Prime Minister in the 1980s: that he used his outstanding skill and wit to keep the ship afloat, inspiring people when he acted on principle but often overseeing policies that he did not believe in. Good intentions were deferred, other people's agendas got priority. Helen Clark is at risk of doing the same and of becoming a casualty of confrontational politics.

The trouble is that political management works, at least in the short-term. In the last week before Christmas that year, the same week as Hobbs' inconspicuous seed announcement, one commentator after another talked about how well Clark had handled the year, how maybe business had 'tamed' her but how she also had 'charmed' it. The *New Zealand Herald* named her politician of the year.

The contaminated seeds were only one issue among many in the first term of Clark's Government. But I believe that how leaders conduct their affairs have effects far beyond the actual decisions. In various speeches that year, Helen Clark had talked about the damaged state of democracy in New Zealand.

> *From the mid 1980s many New Zealanders felt cheated by the political process. Too often they felt that they got the opposite of what they voted for. In all truth they often did. By 1999 trust and confidence in our democratic process was at a low ebb.*[73]

She said that a key task for her Government was to 'restore public faith in the democratic process....It is impossible to overstate just how jaded and cynical many New Zealanders had become about the political process.'[74]

Helen Clark was talking primarily about broken promises. But secrecy, deceptiveness and political manipulation all damage public trust and weaken the fibre of democratic society. The long-term survival of democratic societies, especially in this era of business power, relies on individuals involved in public business – from public servants, to PR people, to the Prime Minister – valuing democracy. The handling of the seeds issue was a classic case of a government abusing power for reasons of short-term political convenience, and in doing so running down the democratic stocks of the country.

Governments like the Labour-Alliance coalition talk as though their greatest mission is to prove that a 'centre-left' government can survive two or three electoral terms. The now familiar formula – Helen Clark has studied the methods of Tony Blair closely – is to be as un-left as electorally possible and run a government of political managers and PR people. My view is that the mission, and the hope, of centre-left governments is the strengthening of democratic government: a renewed and more independent public service, freedom of information, resources put into improving state news services, more open governmental processes and, most difficult, reaffirming that, while businesses can have their say, it is the citizens who decide the policies in a democracy.

Like many political issues today, all the various weaknesses of the country's democratic infrastructure combined to determine the way the contaminated crops incident was handled. There was a lack of willingness by senior public servants to challenge the Prime Minister (even when she was allowing something possibly illegal to go on). The secrecy precluded alternative advice and covered up the poor decision making. The news media were effectively managed so as to render their coverage feeble. There was closed and centralised governmental decision-making, from which even Cabinet could be sidelined.

The unlawfulness of the Government's actions seems clear. The Hazardous Substances and New Organisms (HSNO) Act does not permit any new organism to be imported or possessed in New Zealand. Full stop. The 0.5% contamination thresholds agreed to by Helen Clark and rubber stamped by Cabinet did not change the HSNO Act. It continued to be illegal to import or possess even one GE seed unless a formal application had been approved by the Environmental Risk Management Authority (ERMA). If the Government wanted to legitimise the contaminated crops, there had to be an application to ERMA for release of a new organism into the environment. There was not and so, by introducing the import thresholds and allowing the contaminated crops to stay in the ground, the Government was probably agreeing to illegal activities.

There could easily have been alternative advice if the Government had been open to it, as we saw from the highly critical report written by Lindie Nelson and Oliver Sutherland. Here were the deputy chair and another member of the ERMA board challenging the cursory way in which environmental impacts had been dismissed. They were unhappy that the known GE plants were just being left in the ground. They could see no justification for the needlessly high 0.5% tolerance level for GE contamination in crops, nor for allowing any contamination at all if testing showed it was there. As they pointed out, the Government had adopted a 0.5% threshold 'for no apparent reason', which potentially allowed 13 million GE sweet corn and maize plants to be grown in New Zealand every year yet still be defined as GE-free.

Where the officials had carefully only talked about very small sounding percentages of contamination (parts of a percent), Nelson and Sutherland talked about how many plants in the environment those small percentages actually meant. If the whole 5.6 tonnes of contaminated seeds were eventually planted out, the most conservative contamination estimate (Dr Hannah's 0.04%) amounted to over 15,000 'completely uncontrolled GM plants.'[75] If Nelson and

Sutherland's own estimate was right, the number would be over 30,000. Either way, it was by far the biggest known release of GE plants ever into the New Zealand environment.

Did none of the public servants say this? Here were senior officials from the main government organisations responsible for environmental protection. When Helen Clark said the crops must be pulled out, they studiously set about implementing her wishes. But when she changed her mind, did they challenge her that the new course of action was wrong? As one person close to the issue put it: 'The officials felt intense peer pressure....Individuals dive for cover – everyone dives under the covers together – to avoid having to take a stand.'

Part of the problem was the way the public servants were required to keep everything about the incident secret. New Zealand has not had an official secrets act for twenty years. What possible grounds

There is a patchwork of cornfields between the small, predominantly Maori settlements like this on the plains around Gisborne. Yet Maori concern about genetically engineered food did not figure in the decision-making.

could there be for expecting public servants to keep the Government's secrets about a less than honourable decision? They should not have had to.

I have left until last the biggest reason why it was important to Helen Clark to play down and hide the contaminated crops incident. She had come to office assuring the country that Government decisions on genetic engineering were on hold while the whole issue was subjected to rigorous examination by a Royal Commission. Then, right in the middle of the Royal Commission, news of the contaminated sweet corn arrived.

The Royal Commission on Genetic Modification had been set up by the Labour-Alliance Government in part as a sop to the Green Party and also as a way of keeping a lid on such a volatile issue. Throughout the year 2000, the public, protest groups and all manner of interested organisations had been pouring their efforts into making submissions to the Royal Commission instead of pressuring the Government. The last thing Helen Clark wanted was a controversy that inflamed the GE issue right in the middle of the Royal Commission. I suspect this was the main reason why Clark decided to keep news of the contaminated crops from the public.

But there was also the question of whether to tell the Royal Commission that, at the same time as it was conducting hearings and collecting evidence around the country, thousands of GE corn plants were being left to grow freely – in three cases, just down the road from where the Commissioners were holding workshops and hui as part of their Maori consultation.[76]

A Royal Commission is a very strict, statutory process conducted under a warrant issued by the Governor General. Could the Government withhold such a major piece of information from its own Royal Commission? If it did withhold the information, it was compromising the commission process and, in part, invalidating its conclusions. But if it provided the information, it would undermine the picture being painted for the Commission that there was already a strict and trustworthy system of controlling genetic engineering in place in New Zealand.

The solution put to Cabinet, as we saw earlier, was to present a 'careful and balanced'[77] report on the contaminated seed issue to the Royal Commission. These were the same words that were used about informing the public of the 'full dimensions of the issue', which, as we now know, was *very* careful and balanced.

The Government's sensitivity about how the contaminated crops might impact on the Royal Commission can be seen in way Ministers carefully controlled and politically vetted what the Commissioners were told.

The Commissioners had been given some kind of reassuring verbal briefing earlier, but it was a full two months after the Cabinet decision before a written report to the Commission was finalised. The Government had insisted that the paper be cleared by a Cabinet committee before being sent to the Royal Commission. It was submitted to the Cabinet Finance, Infrastructure and Environment Committee for approval on Wednesday, 28 February 2001.

This was very convenient timing. The Royal Commission hearings had started before the contaminated crops were first discovered, but now it was the last full week of hearings before the Commission disappeared from public view to write its report. By leaving the report so long, there would be almost no time left for the Commissioners to raise questions about the incident or for other groups involved in the process to hear about it and question the Government's version of the facts.

The Cabinet committee, attended by eight of the most senior Labour and Alliance Ministers,[78] approved a five-page report on seed contamination and a covering letter to be sent to the Commission. The report, while containing more information than had been given to Cabinet, presented the same conclusions, starting with the claim that the tests had been 'inconclusive': 'Tests could not confirm

The Rt Hon Sir Thomas Eichelbaum GBE
Chair of the Royal Commission on Genetic Modification
Dalmuir House
114 The Terrace
Wellington

Dear Sir Thomas

As you are aware, last November a seed importing company informed the Government that they suspected a shipment of sweet corn seed from the United States may have been contaminated with genetically modified (GM) seeds. Tests could not confirm whether or not GM material was present, but suggested that if contamination had indeed taken place, it was at a very low level.

Based on a preliminary assessment the government considered the risks to human health to be negligible and the risks to the environment to be very low, and took no further action in this case. However, it was clear that action for the longer term was desirable, and the government initiated some new voluntary measures for imported sweet corn seed to reduce the chance that a similar incident could not happen again.

Cabinet also concluded that this matter raised issues that related to the Royal Commissions warrant. Accordingly officials were directed to draft a report setting these matters out. I am forwarding this report to you for the Commission's information. It outlines the facts of the sweet corn case, the legal situation in New Zealand, the Government's new measures, and some of the particular issues raised.

You will appreciate that the Government considered this issue to be urgent, and has therefore introduced these measures immediately. However, the measures do not represent any change in policy and are entirely consistent with existing procedures under the Hazardous Substances and New Organisms Act and Biosecurity Act for dealing with risks related to the introduction, without prior assessment and approval, of new organisms.

Yours sincerely

Marian L. Hobbs

Hon Marian L Hobbs
MINISTER FOR THE ENVIRONMENT AND MINISTER OF BIOSECURITY

Senior Ministers ignored advice from the head of the Environmental Risk Management Authority and sent a letter to the Royal Commission that he had warned was 'misleading'. They clearly did not want the contaminated crops to become an issue during the Royal Commission.

whether or not GM material was present'. Based on a 'preliminary assessment', it went on, the Government considered the risks to human health to be 'negligible' and the risks to the environment 'very low' and so 'it took no further action in this case'. The report also noted that an interim testing regime (no contamination greater than 0.5%) had been introduced and that this was 'very stringent'. It was 'possible, but very much less likely', it said, that testing could detect contamination below 0.5%.[79]

In other words: the contaminated corn shipment was a false alarm, there was nothing to worry about, it was all under control.

But even within the small circle of people who knew about the issue, things were not as uncontentious as the paper suggested. The day before the Cabinet committee meeting, the acting Minister for the Environment, Pete Hodgson, received an 'urgent in confidence' letter from ERMA expressing concern over what the Royal Commission was to be told. The ERMA board had by then met to discuss Lindie Nelson and Oliver Sutherland's memo and expressed its disquiet about the decisions taken. The letter, from ERMA chief executive, Bas Walker, began by saying that the briefing paper to the Royal Commission, prepared by various agencies, is 'the kind of compromise that is inevitable when different interests and views have to be reconciled':

> *I am nevertheless concerned that the material glosses over some very important aspects of the policy position adopted. I note that this has already been the subject of forthright discussion within the Authority which is why I feel it necessary to write this briefing note.*
>
> *My concern has been heightened by the following phrase in the draft letter to Sir Thomas:*
>
> *'... the measures do not represent any change in policy and are entirely consistent with existing procedures under the Hazardous Substances and New Organisms Act and Biosecurity*

> *Act for dealing with risks related to the introduction, without prior assessment and approval, of new organisms.'*

He stated, very bluntly for a public servant, that 'this statement could be seen as misleading....There is no way that [the 0.5% contamination allowance] can be construed as simply a continuation of existing policy. It presents a shift in policy, which (in my opinion) will probably require legislative change to formalise.'[80] Walker was stating that the decision to 'tolerate' contamination up to 0.5% went against the existing law and that the Royal Commission should be told.

Surprisingly, Pete Hodgson and the other Ministers ignored his advice and approved the prepared letter and briefing paper. They were sent to the Royal Commission without alteration. It seems that the wish to keep contentious or unsettling issues away from the Commission won the day.

This means that the Royal Commission was misled by the Government both on the existence of the contaminated sweet corn crops and also on the lawfulness of the contamination threshold (which had been used to legitimise those crops). Between these two deceptions, and withholding all information that did not fit the story, the contaminated crops were successfully kept from becoming an issue during the Royal Commission.

The Commissioners had just one chance to discuss and probe for themselves the issues raised in the Government briefing paper. The second to last organisation presenting its evidence at the hearings was ERMA, on 1 March 2001, the day after the Cabinet committee approved the briefing paper. Appearing before the Commission, Bas Walker raised the 'misleading' advice about the legality of the contamination threshold that he had urgently approached the Government about two days earlier – but in such a muted way that the Commissioners did not appear to notice. Presumably not wishing to contradict the Government, he merely said that 'the statutory basis for [the contamination] protocols is unclear'.[81]

Oliver Sutherland appeared for the ERMA board but also avoided contradicting the Government. He raised the subject of 'possible' sweet corn contamination and, speaking in purely hypothetical terms, discussed how many thousands or millions of GE seeds this might mean in the environment:

> *In respect of sweet corn and the risk to the environment, we don't know the answer to that because an indepth risk analysis hasn't been undertaken for contaminated sweet corn. And, as far as public acceptance of this, we still don't have any answer to that because there's been no public discussion on this newly proposed regulatory scheme.*[82]

Seed contamination was by then a controversial issue in various other countries. It was an issue about which you would expect the Commissioners to be alert. But no questions were asked and the issue passed without comment. While it seems clear that the Government deliberately withheld information from the Royal Commission, and the ERMA witnesses felt reluctant to contradict the Government, I do not understand why the Commissioners did not want to pursue the issue. They accepted, without question, the calm assurance that a possible release of GE plants into the environment had turned out to be nothing.

The Royal Commission's final report gave the whole issue only a passing mention. It said: 'Government is aware some imported seeds may contain a small proportion of genetically modified contaminants and has therefore stepped up security measures to include testing of all imported seed.'[83] One sentence, bestowing on the contamination testing regime the description 'security measures'. In the whole report there was no mention of the sweet corn incident, the contamination thresholds, the legality of the 'security measures', how many plants 'a small proportion' of contaminants meant or whether that mattered.

It seems that the Royal Commission had simply accepted the Government briefing paper without any independent thought. The Commissioners had uncritically trusted the advice provided by the people in authority in the same way that, in their conclusions on genetic engineering, they effectively said that the public should do as well.

The Royal Commission, in its warrant, had been empowered to publish all the evidence it received, unless in some case it thought it 'proper' to withhold something. All submissions, extra evidence, witness briefs and hearing transcripts were placed on the Commission website. But, strangely, the letter and briefing paper from the Government on the contaminated sweet corn were not made public. Wittingly or unwittingly, the Commission co-operated in the Government's strategy of keeping all information about the issue from the public.

On top of misleading the Cabinet and keeping the public in the dark, we must now add compromising the Royal Commission process – all for the sake of accepting the solution most favourable to a multinational seed company. We are left wondering why Helen Clark did not listen to her first instincts and simply order that the contaminated crops be removed.

Part of the responsibility rests with the officials, whose advice slid from noting evidence that 'can hardly be ignored' to talk of 'negligible traces' of contamination. But even if some Ministers were inclined to believe the shifting advice, they should have questioned, checked and rechecked the facts and evaluations they were being given. This was a serious issue. The trouble was that it suited them to have the issue go away. Yet if they were really comfortable with their decisions, why did they go to such efforts to hide them?

Although the story involves New Zealand politicians, New Zealand officials and a New Zealand Royal Commission, and it is

The Royal Commission was misled by the Government both on the existence of the contaminated sweet corn crops and also on the lawfulness of the contamination threshold (which had been used to legitimise those crops). All information that did not fit the story was withheld, undermining the Royal Commission process. (Photo: Copyright The Dominion)

they who should be making decisions affecting New Zealand, my view is that the main influence was really the multinational company, Novartis. GE seed contamination, and lobbying to avoid commercially-undesirable controls, were occurring world-wide. In some countries the Governments were standing up to the multinational seed companies, in others they were not. The decisions made by our Government in December 2000 were a small victory for the multinational in a contest going on in many countries.

Subsequent events highlight how expedient and unjustifiable our Government's decisions were. Remember how many times officials and Ministers argued that contamination was inevitable, that it had to be tolerated and that 0.5% was a reasonable threshold level? And that stopping contamination below 0.5% was not really practical, indeed not even technically possible? Remember how contamination below 0.5% was talked of as if it was so 'negligible' and insignificant that it was not an issue?

Just a year and a half after the events in this book, all that changed. With the problem of the contaminated corn crop safely in the past, the Ministry of Agriculture and Forestry conducted a review of border control for genetically engineered seeds, to see if the 'interim' contamination standards were adequate. The May 2002 discussion paper said: 'The central question is whether the [interim system] provides sufficient assurance that unauthorised GM seeds are not being imported, or whether the Government should seek greater assurances at the border':

> *[An] option being discussed internationally is to allow the unintended presence of low levels of GM seeds below a certain threshold. This approach is not feasible ... because it would allow GM seeds to enter New Zealand even if they were detected, if they were at levels below the threshold. This would undermine the approval process for GM organisms.*[84]

The threshold option is not feasible. It would allow GM seeds to enter New Zealand even if they were detected. This would undermine the approval process for GM organisms. These are all points that had been ignored or vigorously denied only 18 months earlier.

The discussion paper went on: 'MAF proposes zero tolerance for shipments where GM seeds are detected, but recognises that border control can never be perfect and some concentrations of GM seeds may remain undetected.' In other words, the 'zero tolerance within technical limits' policy that various European countries had been adopting was suddenly practical for New Zealand, too. No one was saying 'If the standard is zero contamination then agriculture based on imported seed might have to virtually close down, with huge economic implications', as Marian Hobbs was advised after the first meeting with industry on 24 November 2000.[85] No one said the country would lose $100 million, or that foreign seed companies might refuse to supply seeds to New Zealand.

The 2002 MAF report also said 'PCR testing is so accurate that it can detect single genes but, in practice, the limit of detection is between 0.01% and 0.1% because of the limits on the size of the sample'. That is 10-50 times more sensitive than the alleged 'as-accurate-as-we-can-get' 0.5% threshold. Once sampling limitations were taken into account, the report said at least 0.1% sensitivity was attainable.

Ever since the first meetings between officials and the industry representatives, the zero tolerance option had been strenuously resisted and the 'US standard' pursued. Cabinet was told 0.5% would have to be accepted for 'purely practical reasons'. Marian Hobbs argued publicly that strict testing for GE contamination would mean destroying all the seeds. The Royal Commission was told that 0.5% was 'very stringent'.

But none of this was true. It is hard to escape the conclusion that the 'inevitabilities', the 'only practical options' and 'limits of testing' had all been excuses to justify why nothing needed to be done about the GE contaminated sweet corn crops. There was one season of the threshold regime in 2001, and then MAF began saying exactly the opposite. By May 2002, MAF was saying that accepting 'tolerable' GE contamination was not feasible for protecting the New Zealand environment or for upholding the law.

I do not know what prompted MAF's turn around. I would not be surprised if it was local companies like Cedenco, Heinz Wattie and Talley's (their interests had always been different to Novartis') wanting systems that assured they could get GE-free seeds. It is also possible that the objections of the ERMA board were finally having an effect.

However, in November-December 2000, political expediency ruled. With decision making restricted to a 'partnership' of 'business leadership alongside government leadership', it became possible and necessary to 'reinterpret' and deny the evidence of the contaminated sweet corn crops so that nothing had to be done. The 0.5% threshold could be used to redefine the contamination out of existence. The contaminated crops could be left in the ground and everything kept hidden.

John Ralston Saul spoke about this kind of rational expediency when he was in New Zealand in June 2002. He talked about how, on environmental and public health issues, the 'rational, masculine' approach of people in authority is one of 'can't tell the public, markets would suffer, we don't believe there's a problem.' 'The idea is that you have to lie to the public because you are loyal to the economy,' he said.

Such people talk of 'inevitabilities', he said, so that, divorced from context and ethics, their policies seem logical. He then quoted author Marshall McLuhan, saying that 'nothing is inevitable as long as you pay attention.'[86]

'The most important thing to look for in leaders, whether left or right,' Saul said, 'is that they are ethically anchored'. Overall, he was not arguing 'that self-interest or business is bad; but that they cannot lead a decent society. Society must lead them.'[87]

Society leading means democracy – 'slow, inefficient, uncertain',[88] as we know, but nonetheless providing space for the contributions of many and hopefully cultivating decisions that are sensible and ethical, and not merely 'rational'. The kind of society where, when something goes wrong, the leaders can be trusted not to use secrecy and PR to cover-up the 'crime', leaving it to the public, unsuspectingly, to eat the evidence.

APPENDICES

CHRONOLOGY

10 October 2000	GE contaminated sweet corn seed consignment, from the seed multinational, Novartis, arrives in New Zealand.
16 October 2000	Royal Commission on Genetic Modification begins formal hearings.
17 October 2000	Planting of the contaminated sweet corn seeds begins.
1 November 2000	The food processing company Cedenco receives positive test results of GE contamination.
10 November 2000	MAF and ERMA notified of the contamination by Novartis.
24 November 2000	Further positive contamination test results arrive. Helen Clark says her 'bottom line' is that the crops must be pulled out. First meeting between government officials and representatives of the corn producing companies and Novartis.
29 November 2000	Second meeting between officials and companies/PR people. First advice to Ministers recommending a policy of allowing up to 0.5% GE contamination.
1 December 2000	Public submissions to the Royal Commission close.
4 December 2000	Cabinet passes urgent regulations to allow destruction of crops, but Helen Clark is going off the idea of destroying the crops.
5 December 2000	ERMA analysis of GE sweet corn test results and MAF 'national economic impact' report produced.
8 December 2000	Cabinet paper on issues to Ministers.
11 December 2000	Cabinet accepts the principle of 'tolerable' GE contamination of crops. The plan to destroy the contaminated crops is dropped.
19 December 2000	Marian Hobbs fronts the 'public announcement' that wasn't. ERMA board members first told about contamination.
7 February 2001	Special meeting of ERMA board to discuss GE seed contamination.
28 February 2001	Cabinet Finance, Infrastructure and Environment Committee approves a 'careful and balanced' report to go to Royal Commission.

DOCUMENTS

The following government documents have been included so that readers can judge them for themselves. Also, a set of the media documents have been included as they give a rare insight into how media management is done.

The documents have been retyped to remove identifying marks, but otherwise they are identical to the originals.

1. The first media contingency plan, in case news of the contaminated crops leaked to the news media before the Government was ready: 'Testing reveals low levels of GM seeds in conventional seeds', 1 December 2000.
2. The main Cabinet paper considered by Cabinet on 11 December 2000: 'Possible contamination of imported seed with genetically modified material', 8 December 2000.
3. One page assessment of the environmental implications of having the GE sweet corn in the environment, attached to above Cabinet paper: 'Overview of environmental and health issues of low levels of GM contamination in sweet corn.'
4. Ministry of Agriculture and Forestry 'National economic impact of an imported seed ban', December 2000.
5. Question and Answer sheet for Ministers, 'Sweet corn seeds', 15 December 2000.
6. Marian Hobbs' media statement: 'New safeguards against GM seed introduction', 19 December 2000.
7. The media 'song sheet' prepared for Marian Hobbs: 'Key points for the Minister', 18 December 2000.
8. 'Back pocket notes' for Marian Hobbs' media announcement: 'Suggested responses to questions about sweet corn if asked, undated.
9. ERMA board members Lindie Nelson and Oliver Sutherland's confidential memo to other board members raising concerns about the contaminated corn crops: 'GM contamination of seed', 15 January 2000.

Testing indicates low levels of GM seeds in conventional seeds

Contingency plan for responding to questions (not intended for proactive announcement)

Initial response

Testing by the cropping industry, although inconclusive, has revealed the possibility of very low levels of genetically modified content in a batch of seeds imported into New Zealand, some of which has been planted. The plants are at an early stage of growth. They are not yet at a stage where there is any possibility of affecting other plants, for example through pollen spread. Industry members are working with government agencies to assess the potential risks and review testing protocols. This will be completed before the plants are at a stage where they could affect other plants of the same species. The environmental risk appears to be low.

If the questioner is aware that the seeds in question are sweet corn

There is no risk to human health. While the crop planted was intended for human consumption, it has not reached the stage of being harvested and, therefore, has not entered the food chain. Further plantings have been halted. The corn which might result if the plants were allowed to continue growing is acceptable for human consumption under current requirements of the Australia New Zealand Food Authority (ANZFA).

If asked whether this has happened before

Not as far as we are aware. It is now possible to detect what may not have been detectable previously.

If there are questions about testing of the seed

The seed was previously tested and no evidence of seeds from genetically modified corn was found. More recent tests after some seed was planted have indicated the possibility of a small number of genetically modified seeds being mixed with the conventional corn. However, information about the sampling and testing which has been done indicates that the results are not clear.

If there are questions about how government agencies found out about the seed

The companies involved in importing the seed advised the government agencies of the tests and we are working through these issues together.

If there are questions about the affected crops being destroyed

No decisions are being made until the available test results are fully evaluated, but the situation is being closely monitored. There is plenty of time to make any decision, as the crops are weeks away from pollination.

If there are questions about breaches of the voluntary moratorium

It appears that this may have been an accidental breach of the Hazardous Substances and New Organisms Act of 1996 and of what was intended by the moratorium.

Office of the Minister for the Environment and Minister of Biosecurity

Chair, Cabinet

POSSIBLE CONTAMINATION OF IMPORTED SEED WITH GENETICALLY MODIFIED MATERIAL

Summary

1. A recent approach by industry has brought to light a number of issues with regard to low level inadvertent contamination of imported seed with genetically modified material, as well as gaps in New Zealand's border controls. This is legally significant as, under the Hazardous Substances and New Organisms (HSNO) Act, the possession of new organisms (including genetically modified organisms such as seeds) may be an offence.

2. These issues need to be seen in an international context, which includes an increasing potential for inadvertent contamination of seed by genetically modified (GM) material, and an absence of internationally accepted standards for tolerable levels of GM contamination and testing for GM contamination at importation. The issues also need to be seen in the context of New Zealand's seed usage and seed production businesses and the present Royal Commission and voluntary moratorium.

3. This paper outlines both the steps already taken to deal with these issues, and the medium term approach proposed, particularly to deal with the next growing season. It also invites Ministers to agree that the Royal Commission be briefed on the issues and invited to consider them, and recommends that a clear and carefully balanced public statement about the issues and the actions taken to deal with them be made.

Background

4. Under the Hazardous Substances and New Organisms Act 1996, the possession or release of a genetically modified organism (GMO) is not permitted without an approval from the Environmental Risk Management Authority (ERMA). To date there have been no applications for the release of a GMO in New Zealand, and with the voluntary moratorium in place, none are expected.

5. A recent approach from several seed importers has alerted ERMA and MAF to the issue of contamination of seed with genetically modified material, and to the question of how New Zealand defines 'GM-free'. This brought to light:

- a possible gap in border controls with respect to checking seed imports for possible GM contamination

- the issue of how New Zealand wishes to treat the issue of GM contamination of non-GM imports
- the question of whether New Zealand will allow some tolerance for a low level of inadvertent GM contamination, on what basis and to what extent.

6. This in turn led to the realisation that it was not possible to establish a regime which would provide absolute certainty that a batch of seed is GM free. Absolute certainty would require every seed to be tested, and this would require every seed to be destroyed. Zero risk of the contamination of a shipment of seed with genetically modified seeds is therefore impossible to assure and only theoretically possible to achieve.

Comment

International Context

7. Investigation of the issue of seed contamination revealed the following:

- There is an increasing potential for inadvertent contamination of non- GM seeds with GM material. This is because of the increasing volumes of GM crops grown, and the characteristics of handling and production systems, particularly in countries where there is little requirement to distinguish between GM and non-GM crops. Further, the number of GM-crops commercially available on the international market is increasing, spreading the scope of this problem.

- There is an absence of international standards for either quality assurance, or for border control testing for GM-material, although European countries are in the process of developing such standards.

- Certain seed companies are taking steps to manage the issue of contamination, and have in place quality assurance schemes designed to deliver 'GM-free' products. However, as noted above, even the most rigorous of such schemes cannot assure absolutely that no GM seeds are present where GM crops are commercially available.

New Zealand Context

Agricultural industry

8. New Zealand imports seeds from several countries (e.g. the US and European Union countries) that are currently growing GM-crops. In that respect, it is at risk of importing GM-contaminated seed. The only way of eliminating that risk would be to ban all imports of seeds from these countries.

9. New Zealand also has a substantial business growing seed for use in other countries (e.g. in the northern hemisphere) and as part of the cycle of developing new cultivars of staple crops. This too carries a risk of contamination.

10. Whilst there is no reliable evidence of GM contamination, a preliminary analysis of the impact of preventing the importation of any seed into New Zealand with the

potential to be genetically contaminated, estimates that there would be a loss of the order of $100 million per annum (allowing for the inclusion of seed multiplication industry and home garden seeds). The area of crops that could potentially be subject to very low levels of GM contamination would cover some 40 000 to 50 000 ha. Effects would be regional (in particular Waikato/Bay of Plenty, Poverty Bay, Hawke's Bay and Canterbury), and would be expected to have considerable impact on individual farmers and growers.

11. There are currently a number of crops growing in New Zealand that have GM counterparts commercially available overseas. These include maize, sweet corn, canola (grown in New Zealand for seed), tomatoes and squash.

12. The approach from seed importers appears to have arisen from a specific case of possible seed contamination. The case involves a shipment of sweet corn seed from the US to New Zealand, which was supplied to four companies, and some of which has been planted. Although initial testing suggested contamination at around the 0.5% level, a more detailed evaluation concluded with a high degree of confidence, that if present at all, there is unlikely to be reliably discoverable GM contamination present in the particular shipment. The evaluation also concluded that further sampling and testing is unlikely to change this conclusion.

Regulatory context

13. Section of 109 of the HSNO Act makes it an offence to knowingly import or release a new organism (including a GMO) or to knowingly recklessly or negligently possess a new organism. The Act also provides that an enforcement officer appointed under the Act may issue a compliance order requiring a person to take action necessary to comply with the Act. If a consignment of seed entered the country and was subsequently found to be contaminated with viable GM material, this would represent a likely breach of the law and could be dealt with through a compliance order. Material intercepted at the border can be dealt with under the Biosecurity Act.

14. At present there are no border controls in place for the testing of compliance of seed imports with respect to GM contamination. However existing documentation advises importers that the importation of a new organism (including a GM organism such as seeds) requires an approval from the ERMA.

Public Concern

15. As is the case in many parts of the world, there is a high level of public concern over the issue of genetic modification. The difference in New Zealand is that there is a Royal Commission currently underway. The Commission will report on strategic options for New Zealand in June 2001. Until the Royal Commission has reported there is a voluntary moratorium on most applications to field test and all applications to release a GMO.

16. For these reasons it is necessary to carefully consider the options for action.

Short term steps

17. Consideration needs to be given as to whether New Zealand should take some early steps towards a cautious and interim approach to dealing with this issue, bearing in mind the fact that it is not possible to be certain that no GM-material is present. It needs to be clearly understood that, if this were done, New Zealand would be amongst the international leaders in the setting up of systems for providing assurances that GM contamination in non-GM seed imports does not exceed defined low levels. So far only a small number of countries (e.g. Switzerland, Denmark, Australia, and Japan) have established explicit import standards.

18. In light of the context described above, a possible interim system for inspection is being developed initially for sweet corn, which will then form the basis for similar systems for other crops. The system would provide clear sampling and testing procedures, documentation such as import declarations, and audit checks by the enforcement agency.

19. Ministers are advised that, for purely practical reasons, the system would have to be set up in relation to a very low but nevertheless explicit level of allowable inadvertent contamination. The present interim proposal is for a maximum of 0.5% contamination in sweet corn.

20. In addition, the means to appoint enforcement officers under the HSNO Act (and so issue a compliance order requiring a person to take action necessary to comply with the Act) is now available. On 4 December regulations for qualifications for compliance officers were approved by Cabinet and authorised by the Executive council on 4 December 2000, and gazetted on 7 December making it possible to appoint officers from 8 December. Authorisations must be given on a case-by-case basis.

Action for the current growing season

21. Most sweet corn crops are already planted for the present growing season. The evidence suggests that for both these crops, and the small amount of seed that remains to be planted, if there is any GM material at all, it is not above the level in the proposal referred to in paragraph 19. Further, there appears to be little possibility of any detrimental environmental or public health consequences from this season's crops (see assessment in the letter attached as Annex 1). This conclusion follows from considering:

- The fact that there will be no further major imports of seed for planting this season.

- A detailed evaluation has suggested that there is negligible contamination of the batch of sweet corn seed that gave rise to the original concerns. In addition there is no evidence of contamination of other batches that could be used to complete planting.

22. With respect to crops other than sweet corn currently planted in New Zealand, there is a theoretical possibility of inadvertent contamination. However, there is no evidence to provoke suspicion of such contamination, nor is there a practicable way

of testing for such contamination with planted crops.

Future steps required

23. The potential risks highlighted by recent events require that best practice guidelines for border inspection and compliance need to be developed. These guidelines need to consider two issues: the ability to effectively and practically sample shipments of seed that might be inadvertently contaminated by GM material, and the ability to test such samples for the presence of GM material. Development of such guidelines will need to be done bearing in mind that this may impact on New Zealand agriculture and ultimately on consumers.

24. In the first instance, the interim system for sweet corn referred to in paragraph 18 needs to be developed into a practicable and workable system that can be effectively used for border control and HSNO enforcement. Sampling and testing procedures for key crops other than sweet corn also need to be established.

25. Accordingly MAF and ERMA will develop these interim arrangements into a set of best practice guidelines for the purposes of administering the HSNO and Biosecurity Acts, and will report to Cabinet on the results of this work. As arrangements for seed supply for the new (2001) growing season commence as early as March, this work should ideally be completed by then.

26. Because these interim arrangements will involve low but defined levels against which tests are made, avenues for formalising these levels will need to be investigated. This may be able to be dealt with through HSNO regulations but there might also need to be an amendment to the HSNO and/or Biosecurity Acts.

Communication

27. Because of the sensitivities around the Royal Commission and the voluntary moratorium in New Zealand, communication of the issue, and the steps underway to deal with it, will need to be carefully considered. Both the public and the Royal Commission will need to be informed of the full dimensions of the issue in a clear and balanced way. Accordingly, it is recommended that a statement be made, which carefully outlines the dimensions of this issue, and clearly describes the next steps to be taken in dealing with it.

28. Officials recommend that the Royal Commission will need to be provided with a detailed report of any information that has come to light during the investigation of this issue, and be invited to consider these issues as part of their deliberations.

Consultation

29. The following departments have been consulted and their comments incorporated in this paper: the Ministry of Agriculture and Forestry and Department of Prime Minister and Cabinet. ERMA New Zealand has also been consulted.

Financial Implications

30. There are no direct financial implications from this paper. Any financial implications will be identified in the report back on further work.

Regulatory Impact Statement

31. Not applicable.

Recommendations

32. It is recommended that Cabinet:

a) note that several seed importers voluntarily alerted MAF and ERMA NZ to the issue of potential seed contamination with GM material, and to a possible gap in New Zealand's border controls

b) note that inadvertent GM contamination (at low levels) in seed imports appears to be unavoidable and increasingly likely to occur

c) note that the possession of GM organisms (including plants and seeds) represents a likely breach of the Hazardous Substances and New Organisms Act 1996

d) note that the Royal Commission on Genetic Modification is currently considering the strategic options for New Zealand, and that there is a voluntary moratorium on applications for GMO field trials and releases

e) note that no international standards have been developed for quality assurance or border controls in respect of any risks of GM contamination of seeds and that New Zealand would be amongst the international leaders in the setting up of systems for providing assurances that GM contamination in non-GM seed imports does not exceed defined low levels

f) note that an interim system for inspection is being developed for sweet corn in New Zealand based on obtaining a high level of assurance (99%) that any contamination is less than 0.5%; and that information on present sweet corn seed imports is that, against these parameters, there is no reliable evidence for concern about GM contamination

g) note that ERMA New Zealand advises that environmental and public health consequences are considered to be very low or negligible in the event that there may be low levels of contamination of sweet corn with GM material

h) note that officials (MAF lead) will report back to Cabinet early next year, advising on progress with the setting of best practice guidelines that can be practically used to provide assurances about defined low levels of contamination for border protection and HSNO enforcement purposes, and including any current risks from previously imported seeds

i) agree that the Royal Commission on Genetic Modification be briefed on the background and contents of this paper once approved by Cabinet, and prior to any public announcement that might be made

j) direct officials from MAF and MfE, assisted by ERMA NZ, to prepare for Cabinet approval a report on this issue, to be referred to the Royal Commission for inclusion in their consideration of the terms of reference

k) agree that the Minister for the Environment and Minister of Biosecurity make a statement carefully outlining the dimensions of this issue and describing clearly the next steps to be taken in dealing with it

Hon Marian L Hobbs

Minister for the Environment and Minister of Biosecurity

OVERVIEW OF ENVIRONMENTAL AND HEALTH ISSUES OF LOW LEVELS OF GM CONTAMINATION IN SWEET CORN

Providing the assumption that we are only considering corn plants and that the only GM contaminants is corn that has been commercially released in North America then the environmental and health issues can be summarised as follows:

Environmental

The primary risks that have been identified relate to the dispersal of pollen that may be contaminated with GM pollen. Two risks are:

(a) **GM pollen cross-pollinates non-modified plants off site.** This is a very small risk as the extent of off site cross-pollination, while it happens, only occurs when there are other corn crops in the immediate (200m) vicinity. Beyond that pollen viability is so compromised that cross-pollination is very unlikely. If the GM pollen component of the pollen available for cross pollination is less than 0.5% at the most, then the resulting risk would be very low.

(b) **GM pollen contamination of honey.** The likelihood of this may be somewhat greater (though still very low) than for off-site cross-pollination as the pollen does not need to be still viable to be used by bees. This risk would be more apparent in organically produced honey.

As there are no native plants that can cross with corn there is therefore no risk to native flora. There are other issues that have been raised such as horizontal gene transfer and spread of antibiotic resistance. These have been considered in generic ways by the Authority in considering containment applications and, within the assumption noted above, are not considered to present risks in this situation.

Health

On the assumption that any contamination by genetically modified corn will arise only from commercially released types of sweet corn there are, in various combinations, only three types of modifications, Bt resistance European corn borer, glyphosate resistant and glufosinate resistant corn. Each of these have interim approval by ANZFA while they undertake the full health assessments and food from them is already allowed to be marketed in New Zealand. Based on this the risks to human health are regarded as being negligible. In our opinion, the full assessment if it was to change this conclusion would not do so to a degree that would negate the current advice.

Donald Hannah

1 December 2000

December 2000

ANALYSIS – MINISTRY OF AGRICULTURE AND FORESTY

National economic impact of an imported seed ban

An economic impact study was done on the impact a ban on importing seed with the potential for being genetically mixed would have for a period of two years.

The study found that the national economic impact of a ban on seed importation for a period of two years would be of the order of $200 million (allowing for the inclusion of the seed multiplication industry and home garden seeds).

The regional impacts, however would be significant in the regions of Waikato/Bay of Plenty, Poverty Bay, Hawke's Bay, and Canterbury.

The impact on individual farmers and growers would be even greater, as there is a reasonable number of farming operations which rely heavily on the crops involved.

Farmers would be further affected if crops had to be destroyed, as there are very limited options for further crops to be sown this season. For affected farmers, this would mean they have incurred up-front costs, and they would lose this season's income. Some would face severe financial difficulty.

The pork and poultry industries would also be affected, as much of the maize grown is used in pork and poultry feed. This feed would have to be imported.

The processing sector would also be affected as a significant proportion of maize and most of the sweet corn is grown for processing.

The analysis is based on the assumption that the ban would be on any seed, from any country, with the potential of being genetically mixed, either through deliberate genetic modification, or with the risk of cross contamination.

To put the issue into context, there are approximately 40 million hectares worldwide of commercial plantings of genetically modified crops. The majority of these are in the US, Canada and Argentina. The major crops are corn in the United States, canola in Canada and soya beans in Argentina.

The source country of the exporting seed cannot always be known however, because here is an opportunity for "seed mingling" through spot markets, which may result in an unclear GM status for a seed line.

Imported seeds New Zealand is reliant on, which has the potential of being genetically mixed are maize, sweet corn, soya bean, canola, squash, pumpkin and tomato.

Crops grown in New Zealand which are unlikely to be contaminated either because the seed does not met the criteria (any seed from any country, with the potential for genetic mix), or because New Zealand is not reliant on imported seed are: wheat, barley, oats, grass seeds, clover, horticultural rootstock, cabbages and brassicas, sorghums, rice, Asian vegetable seeds, asparagus, peas, potato, papaya, flax and linseed, carrots, onions, cucumbers and gherkins.

Analysis Used

Crops excluded from the analysis, and an explanation for doing so, are:

The Seed multiplication industry: this is an important industry, largely in Canterbury. Multiplication applies largely to clover and grass seed. The grass seed is mainly for turf culture. The multiplication is for proprietary brands, and there is no data available on the amount of seed imported or the area grown. It could not, therefore be included in the analysis. The practice is high return, and it is believed it could be worth around $30 million per year. The industry is very important to the local economy because of its high returns, and because it is at the cutting edge of the industry and introduces new technologies.

Soya bean: again there are no firm statistics on areas grown. It is known that the area is small.

Cotton: there is no known data, and most cotton seed, which is imported, is used in the feedstuff manufacturing process (largely based on oil).

Seeds for the home garden: this is a major industry with a large proportion of seeds grown for the home market coming from overseas. Again, no accurate figures are available outside the firms directly involved.

Main Crops included in the Exercise:

Crop	Area(ha) (approximates)
Maize (grain)	15, 000
Maize (silage)	20, 000
Sweet corn	6,500
Tomatoes	2,200
Squash	7,500
Canola	1,500

Maize Silage

This is mainly grown as stock feed for dairy cattle, and to a lesser extent for beef cattle (especially feedlot type operations) and sheep. Average gross margin for 1999/00 was $1,300/ha.

Total national cost would be $1,300 x 20,000 ha = $26,000,000/year.

There would also be some cost to the dairy industry due to lower milk production as a result of either not feeding the silage as a supplement, or feeding an alternative feed. This is difficult to calculate as silage can be fed at different times of the year, with different production responses. Assuming an average response of 30 kg silage DM to 1 kg milksolids, the national area would yield 11.7 million kg MS. The assumption is that farmers would grow alternative crops, although there would be some restrictions on area – most maize silage is grown by contractors, and maize would produce 50 - 80% more DM than most alternative crops. The general assumption would be that 50% of the maize would be replaced with alternatives such that the total loss of milksolids would be 5.85 million kg, valued at $4/kg, giving a total loss of $23,400,000/year.

Maize silage is a major component of the feedstock at the 5 Star Feedlot in Canterbury. They would replace the silage, but at a higher cost. No allowance has been made for this extra cost within this estimate.

Maize Grain

The primary end uses for maize grain are as animal feed (mainly to the poultry and pork industries), for starch extraction and for human consumption. Average gross margin for 1999/00 was $670/ha.

Total national cost would be $670 x 15,000 ha = $10,050,000/yr.

This grain would have to be replaced by imports, with an extra cost for the transport. Generally New Zealand grain prices are the world price plus the cost of transport. Average transport costs from Australia are around $45 AU/T, and $40 US/T from North America. Assuming an average yield of 9 Tonne/ha, the total amount needed to replace the grain currently grown in New Zealand would be 135,000 tonnes. Assuming 50% was imported from Australia, and 50% from North America, the overall transport cost would be:

67,500T x $45AU x 1.35 (exchange rate assumption) = 44,101,000
67,500 x $40US x 2.5 (exchange rate assumption) = $6,750,000
Overall total = $10,851,000/yr

Sweet Corn

This is grown for human consumption, either fresh or processed. Area grown in the North Island is 4,700, with an average gross margin of $1,100/ha. Area grown in the South Island is $1,800 ha, with an average gross margin of $1,470/ha.

Overall cost is 4,700 x 1,100 +1,800 x 1,470 = $7,816,000/yr.

Tomatoes

Again grown for human consumption, either fresh or processed. Total area grown is 2,200 ha, with an average gross margin of $2,500/ha.

Total cost is 2,200 x 2,500 = $5,500,000.

Buttercup Squash

This is largely grown for human consumption, with a significant amount exported. The profitability of squash has varied considerably over the last few years, and reliable figures on area and gross margin are difficult to obtain. The assumptions are based on an area of

7,000 ha, and a gross margin of $1,000/ha. The last published GM for squash was $1,600/ha in 1995. However, the profitability of squash has dropped significantly since 1995.

Overall cost is 7,000 x 1,000 = $7,000,000/yr.

Canola

This is an oil seed crop, with an estimated area grown of 1,500ha, and an average gross margin of $1,200/ha.

Overall cost is 1,500 x 1,200 = $1,800,000/yr.

Summary

Total costs would be:

Crop	Cost/year ($M)	Cost over 2 Years ($M)
Maize silage:		
- crop	26.0	52.0
- less milksolids	23.4	46.8
Maize grain:		
- crop	10.05	20.1
- extra transport costs	10.851	21.7
Sweet corn	7.816	15.63
Tomatoes	5.5	11
Butternut Squash	7.0	14
Canola	1.8	3.6
Overall cost	92.4	184.8

DATA SOURCES
MAF Farm Monitoring Reports 2000
Horticulture Facts and Figures 1999
Canterbury Agriculture Gross Margins, September 2000
MAF Personnel

Questions and answers – sweet corn seeds

15 December 2000

1. What prompted the current action on a sampling and testing system?

   Some conventional sweet corn hybrid seeds were imported to New Zealand for the current growing season. Initial testing suggested that the conventional seeds might have traces of genetically modified material. Some of the seeds had been planted and were in the early stages of growth. A detailed evaluation concluded that, if genetically modified material was present at all in that shipment, the amount present was below the level that could be reliably detected. It is also possible that low levels of contamination with bacteria or soil or contact with contaminated equipment may have led to false positive test results.

2. Why is this a concern?

   Importing a genetically modified organism (GMO) without approval is a breach of the Hazardous Substances and New Organisms Act 1996 and the intent of the voluntary moratorium. Approval from the Environmental Risk Management Authority is necessary before genetically modified organisms can be field-tested or released in New Zealand. A voluntary moratorium on applications to the Authority for most field testing and any release of genetically modified organisms is in place until August 2001. In this case, the seed imported was conventional and believed to be GM-free at the time of importing, so the question of an application did not arise.

3. Would the sweet corn grown from this seed be a risk to human or animal health?

   The ANZFA Standard A18, which is the current standard for foods in New Zealand with genetically modified content, allows food from this crop to be sold. This is possible under an interim arrangement which is in effect until the Australia New Zealand Food Authority (ANZFA) has undertaken the safety assessment and approval of certain genetically modified foods. The full application for this particular food, known as Bt11 corn, is still being processed by ANZFA. The interim approval applies while that safety assessment is underway.

4. Are there any risks to New Zealand's environment and biodiversity?

   Any risks are extremely low, given the very low levels of genetically modified material indicated. The crop could cross-pollinate only with other species of corn and maize (*Zeamays*). However, even this is not particularly likely, as corn and maize normally self-pollinate. The pollen is viable for only a short time and travels little distance from the individual plant. There are no native species at risk.

5. Where did the seed come from?

   This consignment of corn seed came from the United States. It was imported for a number of New Zealand companies.

6. What testing was carried out on the sweet corn seeds?

The shipment of sweet corn was 5.7 tonnes. (There are around 20 seeds in 4.7 grams.) 49 000 seeds from this shipment were tested in 12 samples. Of these tests, 10 results have been given to officials in New Zealand. Testing was done by laboratories in New Zealand, Australia and the United States.

Information on the 10 samples was evaluated by ERMA New Zealand scientists and other experts in the field (Crop and Food, and Otago University). Possible GM content was indicated, but not a level that could be reliably detected. Across the batch as a whole, it can be concluded with a high degree of confidence that there is unlikely to be more than 0.04% GM content. This is well below the proposed threshold of 0.5% for sweet corn.

7 Do we know what genetically modified material was indicated?

Tests were for the presence of the Bt gene and for the "markers" commonly used in genetic modification. Some, but not all, of the results showed the possible presence of the Bt gene and others indicated that the "markers" might be present. The Bt gene is a gene from a bacteria that infects and kills insects, and it is used in corn crops to make them resistant to corn borer. When the gene is inserted into the corn plant it becomes resistant to attack from this pest. Bt is widely used in organic agriculture as a biological alternative to chemical pesticides.

8. How could conventional seed become mixed with genetically modified seed?

We do not know in this case. In countries where both genetically modified crops and conventional crops are grown, it is most likely to happen through a batch of conventional seed accidentally having a small number of genetically modified seeds mixed into it during processing, storage, or other activities after the crop is harvested. Even stringent quality procedures make it difficult to avoid low levels of GM content, and this issue is commonly recognised throughout the industry. In other cases it could come about through cross-pollination of part of the crop.

9. Have similar problems happened before?

The Environmental Risk Management Authority and the Ministry of Agriculture and Forestry have no evidence to suggest that there may have been similar problems with earlier batches of this seed imported into New Zealand. This does not mean that it could not have happened in the past. To date there has been no mandatory testing in New Zealand for genetically modified content in seed crops, nursery stock, fresh fruit and vegetables, cut flowers, etc. However, importers are routinely advised that approval under the Hazardous Substances and New Organisms Act is needed for importing genetically modified organisms.

10. Is this situation likely to recur in the future?

Yes it is. As GM crops become more widespread, so the chances of low levels of GM content in conventional crops grow. That is why New Zealand, along with some other countries, is moving to set firm standards so that the issue can be properly managed.

Hon Marian Hobbs
Minister for the Environment and Biosecurity

Embargoed to 9am 19 December Media Statement

NEW SAFEGUARDS AGAINST GM SEED INTRODUCTION

New measures are being developed by the Government to guard against the inadvertent introduction of genetically-modified seed, the Minister for the Environment and Biosecurity, Marian Hobbs, announced today.

Officials are working on a seed testing regime and best practice guidelines to provide backup to the Hazardous Substances and New Organisms Act (which controls the deliberate introduction of new species) and the Biosecurity Act. They will also submit a special report to the Royal Commission on Genetic Modification.

"New Zealand imports seed from a number of countries where GM crops are grown," Marian Hobbs said. "Seeds which could be genetically modified include maize, sweet corn, squash, tomatoes and canola. We feel the need to strengthen border controls with respect to checking seed imports for possible GM presence."

Ms Hobbs said a number of seed importers are taking a very responsible approach to managing the risks of possible GM content in batches of seeds. They have in place quality assurance schemes designed to minimise the risk of GM content inadvertently getting into their products.

They recently alerted the Environmental Risk Management Authority and the Ministry of Agriculture and Forestry to the possibility that some imported conventional sweetcorn seed might contain traces of genetically modified material.

"Although initial testing suggested there might be minute traces of GM content, a more detailed evaluation concluded with a high degree of confidence that, if present at all, the GM material was at levels below that which can be reliably detected, " she said.

Our best advice is that it is not possible to establish a testing regime which would provide absolute certainty that a batch of seed is GM-free. Absolute certainty would require every seed to be tested and this would require every seed to be ground up to extract the genetic material for the test. It is therefore impossible to assure zero risk of contamination of a shipment of seed."

The only way to possibly eliminate the risk would be to ban all seed imports from all of the countries currently growing GM crops, a list which includes many of New Zealand's major trading partners.
"This would result in losses of an estimated $100 million a year affecting farmers and growers in many parts of the country," Marian Hobbs said.

International standards are still being developed for quality assurance or border controls covering risks of GM contamination of seeds.

"New Zealand is taking a cautious approach in establishing systems which will provide reasonable assurances for non-GM seed imports," the Minister said.
"Inspection systems will be developed to enable biosecurity officers and HSNO enforcement officers to consistently inspect imported seeds, based on obtaining a high level of assurance that any inadvertent GM contamination is near the limits of reliable detection.

"We recognise this is a conservative approach and we will be looking to the Royal Commission for advice in its findings on how such issues should be treated in the future."

The new arrangements should be completed by next March when seed supply arrangements for the new growing season commence.

Media inquiries:
Trevor Henry
Press Secretary

Key Points for the Minister

The issue

- No evidence of GM crops growing in NZ
- Problem containable at present
- Major trading partners grow both GM and conventional crops
- Key species are maize, sweet corn, squash, tomatoes , canola
- 100% GM-free is impossible to guarantee for some crop seeds
- Commercial GM crops on international market increasing
- Very low levels of GM can come in inadvertently during storage/handling or through cross-pollination
- Banning imports of seeds with potential to be GM could cost NZ $100 million a year
- Realistic to test only samples
- Testing every seed means grinding up and destroying ability to grow
- Problem is similar to other biosecurity issues (snakes, insects)
- NZ not GM-free. . . GM research for last 12 years: food in supermarkets has GM content
- ANZFA standard A18 applies to food produced using gene technology

Sampling and testing

- Border control system of sampling and testing being established
- Interim proposals are near limits of reliable detection
- Interim standards for all crop seeds to be completed by March 2001
- Interim proposal is 0.5% tolerance for GM content in maize and sweet corn
- Other countries starting to set standards that allow traces of GM content
- Seeds not meeting standard would be re-exported or destroyed
- Need to allow for very low levels of accidental contamination
- Major importers have quality assurance systems
- ERMA was aware of potential risks and advised MAF
- No mandatory testing to date
- Risks are growing - more GM crops are being grown overseas.

Royal Commission and moratorium

- Commission has been briefed
- Does not pre-empt work of Royal Commission
- Commission asked to consider this issue
- Commission will report mid-2001
- Border control standards will be reviewed on basis of recommendations
- Moratorium still applies - any GM in seed imports is accidental

Suggested responses to questions about sweet corn if asked

What has happened to the plants and seeds?
Nothing. There is no risk to people's health or the environment. If the plants and seeds have any GM content at all, it is below the level proposed in the interim standard, which allows for a maximum level of accidental contamination of 0.05%.

Where are these crops growing?
1. In a number of regions of New Zealand. I suggest you talk to the industry spokesperson, Peter Silcock.
2. There are plantings in East Coast, Hawke's Bay and Marlborough.
3. I do not know the actual sites where these crops are growing.

Will we be eating this corn?
If there is any GM content at all, the sweet corn grown from this seed will meet the ANZFA Standard A18, which is the current standard for foods in New Zealand with genetically modified content. The full application for this particular food, known at Bt11 corn, is being processed by ANZFA and an interim approval is in place.

Who imported the seed?
1. I see no need to single out particular companies, especially as they have been acting responsibly. I suggest you talk to the industry spokesperson, Peter Silcock.
2. It was imported by Novartis for a number of New Zealand companies.

What was the sweet corn?
Novartis Jubilee Sweet corn Lot NC9114.

Will there be any prosecutions?
No. Any GM content that was in the batch of seeds got there accidentally. There was no intention to import a genetically modified organism. The companies concerned have acted responsibly in bringing a potential problem to the attention of ERMA New Zealand and MAF.

Will pollen from these crops contaminate other plants?
Maize and corn is generally self pollinating, so the pollen does not spread far and is viable for only a short time. The crop could cross-pollinate only with other species of maize and corn.

Can we get a photo/film of these crops?
I suggest you talk to the industry spokesperson, Peter Silcock.

Fax to: **Jill White**
John Maasland
Lindie Nelson
Tony Haggerty
Colin Mantell
Helen Hughes
Prue Kapua
Leatrice Welsh
Bas Walker

Confidential Memo

From: **Lindie Nelson and Oliver Sutherland**

Subject: **GM Contamination of seed**

Date: 15 January 2001

Kia ora koutou

Since our most recent Board meeting (November 2000), major issues have arisen regarding the importation of GM-contaminated seed to New Zealand. These have led to various actions being taken by Government and by officials. While the Authority members have largely been kept informed with on-going developments (see memos from Bas of 19 December and 21 December, particularly) we have become increasingly concerned at the fact that some major decisions have been made (leading, amongst other things, to the planting of contaminated sweet corn) which have had no active input from the Authority. Of course, the time of year at which all this has occurred hasn't helped.

In any case, it has prompted both of us to look more closely at the issues – Oliver, from the point of his appearance as a witness before the Royal Commission, re the general issue of the acceptance of a specified level of GM contamination of seed and Lindie from the more specific matter of the 5.6 tonne consignment of Novartis sweet corn that kicked off the whole matter. We have been helped in this (i) by Kevin Currie, who has provided us with background information that had not been copied to the Authority to date and (ii) by some input from professional biometricians. All of this increases our sense of unease at the Authority's lack of active input into policies as they have developed.

Background

In November of last year, a 5.6 tonne consignment of Novartis sweet corn seed from the US arrived in NZ and was determined to have some contamination with GM seed. This has prompted two outcomes to date:

(i) The Government has announced an interim proposal for a maximum acceptance level of 0.5% GM content in maize and sweet corn seed, and officials are developing a testing regime which will detect with 99% confidence that contamination is no greater that 0.5%.

(ii) A range of tests (x 10) was performed on the shipment of Novartis sweet corn and an analysis of the results was undertaken by Donald Hannah. He concluded that the seed..... "has less than 0.5% GM contamination...." And so was below the Government's maximum acceptance level of 0.5%. Accordingly, approval was given (by officials/Minister) for the seed, which we understand had been planted at that stage, to remain in the soil. The plants are now maturing at known locations. It is still not too late for the plants to be destroyed before pollen is shed.

(i) **Maximum acceptance level of 0.5% contamination**

The interim approach ("standard") proposed is to establish sampling and testing procedures designed to detect with 99% confidence, that contamination is no greater than 0.5% (see Bas's memo of 21 December).[1]

The agreed approach is a de facto decision to allow release of grain with a maximum of 0.5% GM content. It is a very significant step. Our calculations suggest that for maize and sweet corn, this could translate to releasing up to 13.3 million GM plants per year throughout the NZ planting areas.[2]

We understand and support the practical need to define a standard, but are concerned about the apparent sidelining of the Authority and HSNO (the body and legislation set up to approve such releases) and the incongruity of strict risk evaluations that we undertake for field trial applications for contained growth of GM plants compared to 'back of the envelope' evaluations for this release. We are concerned at the apparent absence of any ex ante in-depth risk analysis to support the 0.5% acceptance level. Moreover, that this contamination level may be construed as being 'GE–free'.[3]

The moratorium is a voluntary one. No consideration seems to have been given to making an application to release GM contaminated seed, including the use of the Ministerial call in powers. Such an approach would allow the full HSNO evaluation process to be triggered.

The Authority has been very cautious in its approvals of field trial containment applications. We have agonised over the risks of very small pollen escapes and how we could prevent these. It is ironic to find that Cabinet and officials are taking a less cautious approach to a release decision. Further, given that PCR can reliably detect GM contamination down to 0.1%, why adopt the 0.5% level (there may be valid reasons but these are not apparent.)

Finally, the "standard" could be used either as a "threshold" criterion or as a "yes/no" criterion. The documents provided so far suggest that intent is to apply the standard as a threshold.

---

[1] See appended analysis (attachment 1)

[2] See appended date (attachment 2)

[3] See appended radio news transcripts (attachment 4)

criterion. That is, if contamination is detected, but it is below 0.5%, it is acceptable. The alternative is to say that if the tests provide positive evidence of contamination, the seeds are unacceptable regardless of the level of contamination. In this alternative case, the release would be less because seeds known to be contaminated would not be planted. There does not seem to have been any discussion of the implications of the alternative application of the standard.

(ii) **Novartis sweet corn**

Once it became apparent that the 5.6 tonne consignment of Novartis 'Jubilee' sweet corn contained some GM contamination, a total of 10 tests were undertaken in 3 or 4 laboratories (NZ, Australia, US) to determine the level of contamination. The results of these tests have been analysed by Donald Hannah who noted various discrepancies and short-comings in the tests and reached:

> "... two, not necessarily mutually exclusive, conclusions:
>
> - If there is contamination by genetically modified sweet corn it is at a level well below 0.5% and is probably not evenly distributed in the sample.
>
> - Taking further samples and analysing them in accordance with the proposed protocol is unlikely to provide significantly improved or certain information over what already exists.
>
> Therefore, with a considerable degree of confidence it can be concluded the Novartis sweet corn ... has less than 0.5% GM contamination and hence, judged by that standard, does not contain a new organism."

As Donald stresses, there were various discrepancies and/or short comings in the 10 tests performed on the seed. (We have studied Donald's report and share his unease at the rigour of the testing and hence the validity of the overall assessment [4]) Nevertheless, we understand that most of the seed was planted and remains in the ground. If we assume that 5 tonnes has been planted, then this is roughly equal to 35,000,000 seeds (there are c.7,000 seeds/kg):

0.04% contamination equals 14,000 plants
0.08% contamination equals 28,000 plants
0.5% contamination equals 175,000 plants

Whatever the figure is, it is a very substantial number of completely uncontrolled GM plants now growing in the field. (NB. By comparison, the 0.4 hectare Pioneer transgenic maize field trial that ERMA approved in 1999 (but which wasn't planted) would have produced a total of just 1450 GM plants – all pollen tassels bagged to prevent pollen escape together with a range of other controls.

We make several observations

(i) since we know that the Novartis sweet corn is contaminated, is it acceptable that it should remain in the ground?

(ii) should we, the Authority established to determine and assess environmental risks, be included in the decisions surrounding this seed/these plants?

(iii) should we "assert our Authority" and step in at this stage, unsure, as we are, of who 'allowed' the Novartis plants to remain in the soil?

(iv) should the public, who we know care about 'genetic pollution' be informed about the whole Novartis sweet corn issue?

(v) is the standard being developed in a way that minimises contamination?

We have discussed this matter with Jill who believes that a special meeting of the Authority should be convened as soon as possible to discuss these issues. Feasible dates for us and Jill are the mornings of 25 or 26 January or the afternoon of 7 February. If we wait until our February Board meeting, the Novartis corn plants may well be flowering.

Regards

Lindie and Oliver

Comments on report by D. Hannah, ERMA

1. 10 samples were taken, of which the integrity of 6 was called into question – this level of question about the data is of concern.
2. Of the 6 questionable samples 4 were noted as being taken from sacks that had been opened in the field. It is unclear exactly what the "field" means, and thus it is difficult to know to what degree there was risk of the sample integrity being compromised e.g. were the open sacks left standing in a field of corn and exposed to pollen, were the sacks opened in a warehouse in NZ and thus had a much lower likelihood of being exposed to GM contamination? Does "field" literally mean a field, or does it mean outside of containment?
3. A further 4 of the 6 samples of questionable integrity had odd results which indicated possible contamination in the lab (2 of the 6 were both from sacks opened in the field, and may have been lab contaminated). Are these samples which may have been contaminated in the lab from 1, several, or all of the laboratories used to test samples?
4. With the possibility of lab contamination the issue of false positives was raised (results based on false positives could lead ERMA to reject samples which in reality are GM free but at least this is more conservative). Given the rate of lab contamination (4/10) which seems high, it raises the question whether there was an equal risk (through either poor lab practices, or machine detection problems or whatever) that false negatives could occur, which would be a major concern to ERMA who would be at risk of allowing samples which tested negative, but which in reality did contain >0.5% levels of GM, into NZ.
5. Calculation of 0.04% overall contamination – not clear where this figure comes from. Calculating the mean of the samples, assuming 0% for the negatives, gives a value of (0.5% + 0.1% +0.1% + 0.1% +(6 x 0%))/10 = 0.08%.
6. The report concludes that "...contamination... is at a level well below 0.5% and is

probably not evenly distributed in the sample". This last statement is a major concern, as the draft sampling protocol makes a key assumption that "any GM contamination will be randomly distributed throughout the whole of the lot.".

7. Overall (assuming false negatives are not an issue) then the conclusion that there is <0.5% GM contamination is sound, though perhaps it's being rather bold to say it was "with a considerable degree of confidence" that one draws this conclusion.

ENDNOTES

1. Royal Commission of Inquiry into Genetic Modification, hearing transcript, 1 November 2000, pp. 856 and 857.
2. Ibid., pp. 863, 865, 873 and 874.
3. Ibid., p. 877.
4. Ibid., pp. 879 and 918.
5. Greenpeace International, press release, 'Genetically engineered corn contaminates neighbouring crop', Amsterdam/Hamburg, 12 October 1998.
6. Brett Chase, 'Gene-altered crops are trouble in the wind for organic foods', 11 August 1999, reproduced at http://www.biotech-info.net/GE_crop_trouble.html.
7. Austrian Parliament, answers to Parliamentary Questions on 23 July 2001 and 26 March 2002.
8. Hazardous Substances and New Organisms Act 1996, s. 25 (1) (b), s.109(1)(d)(ii) and s.114.
9. The moratorium, although strict, was technically voluntary to avoid having to rewrite the law.
10. Environmental Risk Management Authority decision, T25 and MON810 maize, 21 October 1999.
11. Ibid., pp. 9-10.
12. Bees do not pollinate corn (it is wind dispersed) but they do collect the pollen for food.
13. Heinz Wattie, Cedenco and Talley's now all have policies of providing GE-free products, however only Cedenco was assuring GE free produce (to one customer) in 2000.
14. Norrey Simmons, 'Glittering Prizes', speech to the Public Relations Institute of New Zealand conference, Auckland, 12 May 2000.
15. Marian Hobbs, speech to the Public Relations Institute of New Zealand conference, Auckland, 12 May 2000.
16. Confidential source.
17. *O'Dwyer's PR Daily*, 26 April 2002, quoted at www.prwatch.org.
18. Cedric Allan, 'Ethical Communications' address to the Strategic Communications Conference, Auckland, 30 March 2001.

19. Ministry for the Environment, 'GM Maize Update'.
20. Bas Walker, letter to Marian Hobbs, 24 November 2000.
21. Ibid.
22. Ibid.
23. Ibid.
24. She may well have genuinely believed this to be true.
25. Ministry for the Environment, 'Testing reveals low levels of GM seeds in conventional seeds', 1 December 2000.
26. Bas Walker, op. cit.
27. Ministry for the Environment, confidential question and answer sheet on the incident, 30 November 2000.
28. Donald Hannah, ERMA, 'Evaluation of information received for Novartis Jubilee Sweetcorn Lot NC9114', 5 December 2000.
29. Julie Vorman, Reuters, 'EPA not convinced on StarLink bio-corn safety', 13 November 2000.
30. Donald Hannah, ERMA, op. cit.
31. Ibid.
32. Ibid.
33. Marian Hobbs, Cabinet paper, 'Possible contamination of imported seed with genetically modified material', 8 December 2000.
34. The Austrian example is described further in the next chapter.
35. Marian Hobbs, Cabinet paper, op.cit.
36. David Wansborough, *Border Control for Genetically Modified (GM) Seeds,* MAF Discussion Paper No. 31, May 2002.
37. Marian Hobbs, Cabinet paper, op.cit.
38. Lindie Nelson and Oliver Sutherland, 'Confidential Memo – GM Contamination of seed', 15 January 2001.
39. Ibid.
40. *PR Week,* 5 July 1999 issue.
41. 'ASTA continues collaboration with AgTrade Coalition', American Seed Trade Association website 'What's New', www.amseed.com, 29 May 2002.
42. American Seed Trade Association website homepage, www.amseed.com, 29 May 2002.
43. Marian Hobbs, Cabinet paper, op. cit., pp. 4 and 5.
44. Ibid., p.3.
45. The Hazardous Substances and New Organisms Act says that no unapproved GE organisms can be brought into New Zealand. That is, zero tolerance. However, the law had not anticipated accidental GE seed contamination and so there was no testing regime at that time for imported seeds. All importers were told they would break the law if they imported GE seeds and it was left to them to make a formal application if they wanted to.
46. Reuters, 'Italy refuses seeds containing genetic material', 24 January 2002.

47. The new system proposed by MAF still had weaknesses, such as only one GE test per seed batch being required (despite the Lot NC9114 experience of different parts of the batch having markedly different contamination levels) and only testing one shipment in three. It was a great improvement, but still cut price border control. Ministry of Agriculture and Forestry, MAF Discussion Paper No. 31, 'Border Control for Genetically Modified (GM) Seeds', May 2002.
48. Ministry for the Environment, confidential question and answer sheet on the incident, 18 December 2000, paragraph 6; and Donald Hannah, ERMA, 'Evaluation of information received for Novartis Jubilee Sweetcorn Lot NC9114', 5 December 2000, p.1.
49. Ministry for the Environment, confidential question and answer sheet on the incident, 30 November 2000.
50. Cabinet Office, *Cabinet Manual,* section 3.3, and the accompanying *Step by Step Guide,* section 1.1, www.dpmc.govt.nz.
51. For readers under about 30 years old, Sir Humphrey was a shamelessly manipulative public servant in an old British television programme called 'Yes, Minister', later 'Yes, Prime Minister'.
52. It is for cases like this that the Cabinet Manual instructs that 'good [Cabinet paper] recommendations don't leave any room for doubt about what is being decided'. *Step by Step Guide,* Ibid., paragraph 39.
53. Marian Hobbs, Cabinet Paper, op. cit., pp. 1, 2 and 3.
54. Ibid., p. 3.
55. Ibid., p. 4.
56. Ibid., p. 6.
57. Ibid., pp. 1 and 5.
58. Ibid., p.2.
59. Donald Hannah, 'Overview of environmental and health issues of low levels of GM contamination in sweet corn', Environmental Risk Management Authority report, 1 December 2000.
60. Bas Walker, letter to Marian Hobbs, 1 December 2000.
61. Ibid., p. 2.
62. Ibid., p. 6.
63. Ibid., p. 5.
64. Marian Hobbs, interviewed in Carroll du Chateau ,'GE: Brave New World', *New Zealand Herald,* 6 May 2000.
65. Ministry for the Environment, confidential question and answer sheet on the incident, 30 November 2000, 8 December 2000 and 15 December 2000.
66. The Government was indeed bringing in seed testing where previously there had been none. But, by accepting contamination, it was such a poor regime compared to what was possible (see next chapter) that it was not really good news.
67. Marian Hobbs, interviewed on Radio New Zealand Rural News, 12.36 p.m., 19 December 2000.
68. Green Party, media statement, 'GE standards for seed protects New Zealand's market advantage', 19 December 2000, issued at 10:44 a.m.

69. *Press*, 'Seed imports scrutinised', 21 December 2000.
70. Helen Clark, Massey University Food Awards speech, 9 October 2000.
71. Phil Love, 'Wiping smiles off faces', *Evening Post*, 3 June 2000.
72. Helen Clark, Business Forum speech, 24 October 2000.
73. Helen Clark, Address to Economist Roundtable, 10 August 2000.
74. Helen Clark, Auckland Chamber of Commerce speech, 16 June 2000.
75. Nelson and Sutherland, Ibid., pp. 2-3.
76. The Commissioners held workshops and hui in Picton on 18 January 2001, Gisborne on 2 February 2001 and Hastings on 10 February 2001.
77. Marian Hobbs, Cabinet Paper, op. cit., p. 5.
78. The Cabinet committee meeting was attended by Helen Clark, Michael Cullen, Sandra Lee, Trevor Mallard, Pete Hodgson, Mark Burton, Paul Swain and Mark Gosche.
79. Minister for the Environment, 'Report for Royal Commission on Genetic Modification on Possible Seed Contamination', 22 February 2001.
80. Bas Walker, letter to Pete Hodgson, 27 February 2001.
81. Bas Walker, Royal Commission of Inquiry into Genetic Modification, hearing transcript, 1 March 2001, p. 04428.
82. Oliver Sutherland, Royal Commission of Inquiry into Genetic Modification, hearing transcript, 1 March 2001, pp. 04421-3.
83. Report of the Royal Commission on Genetic Modification, 2001, p. 148.
84. David Wansborough, *Border Control for Genetically Modified (GM) Seeds*, MAF Discussion Paper No. 31, May 2002, p. 4.
85. Bas Walker, letter to Marian Hobbs, 24 November 2002.
86. John Ralston Saul, Wellington Town Hall, 7 June 2002.
87. John Ralston Saul, *On Equilibrium*, Penguin Books, Victoria, 2002.
88. John Ralston Saul, Wellington Town Hall, 7 June 2002.

INDEX